NOVA SCOTIA'S LOST HIGHWAYS

The Early Roads That Shaped the Province

JOAN DAWSON

NIMBUS
PUBLISHING

Nimbus Publishing Limited
PO Box 9166
Halifax, NS B3K 5M8
(902) 455-4286
nimbus.ca

Printed and bound in Canada

Design: John van der Woude
Front cover: Courtesy History collection, Nova Scotia Museum.
Image has been cropped and colourized by publisher.

Library and Archives Canada Cataloguing in Publication

Dawson, Joan, 1932-
Nova Scotia's lost highways : the early roads that shaped the province / Joan Dawson.

Includes bibliographical references and index.
ISBN 978-1-55109-732-9

1. Roads—Nova Scotia—History. 2. Nova Scotia—History.
3. Nova Scotia—Description and travel. I. Title.

FC2312.D39 2009 388.109716'09 C2009-903372-0

Canada
Canada Council for the Arts Conseil des arts du Canada
NOVA SCOTIA Communities, Culture and Heritage

We acknowledge the financial support of the Government of Canada through the Book Publishing Industry Development Program (BPIDP) and the Canada Council, and of the Province of Nova Scotia through the Department of Tourism, Culture and Heritage for our publishing activities.

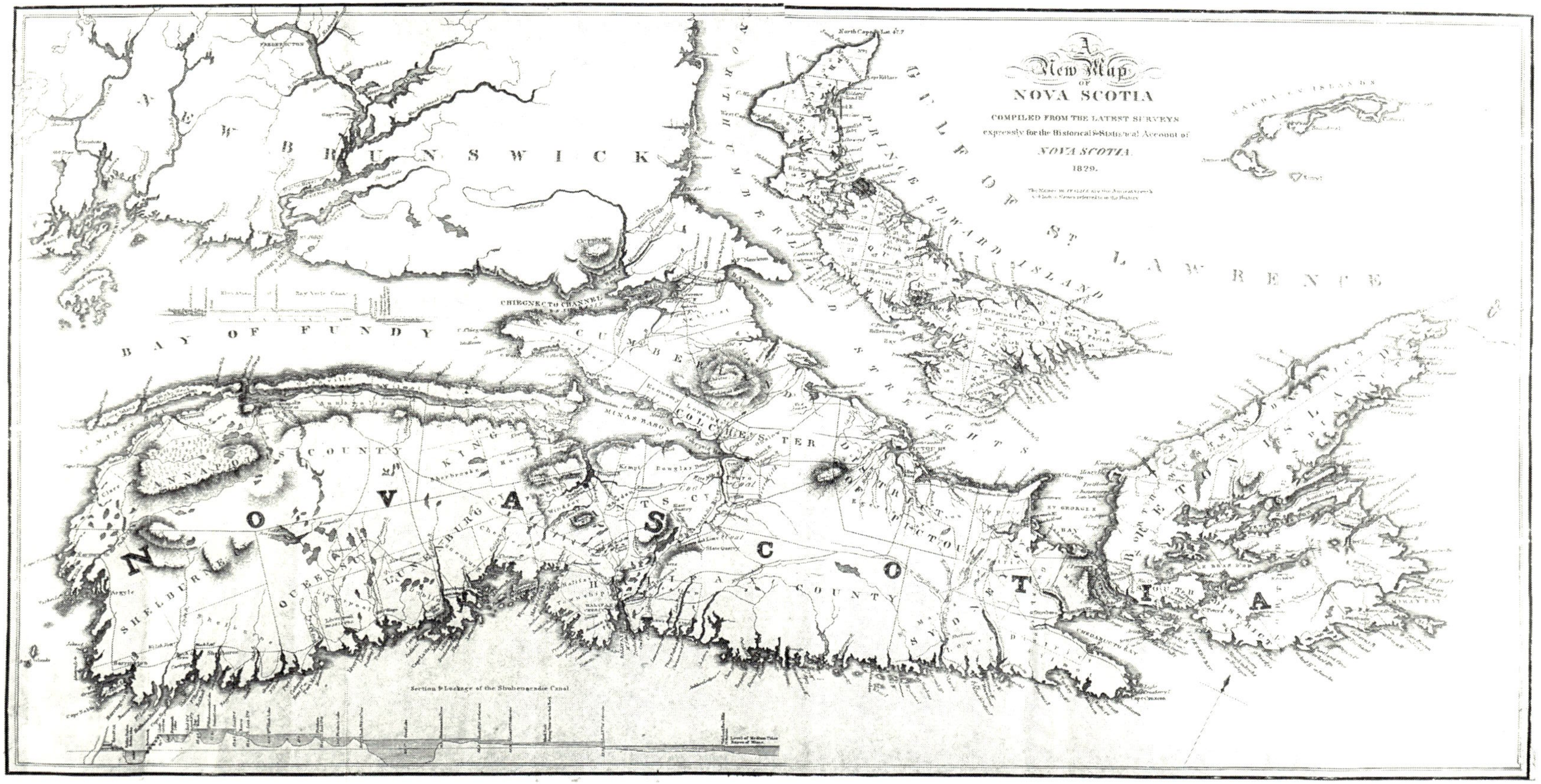

William MacKay. New Map of Nova Scotia compiled from the latest Surveys for the Historical and Statistical Account of Nova Scotia, 1829.

Table of Contents

The road linking Halifax to Truro took longer to develop than the Windsor Road. Truro, like Windsor, had been an Acadian settlement, known by its Mi'kmaq name Cobequid. After the Acadian Expulsion, it was renamed Truro and first settled by Scots-Irish from New Hampshire in the 1760s. Further removed from Halifax, Truro did not develop as quickly as Windsor. It did not have such a strategic military location, nor did it have a university, but it became an important hub of transportation as the province developed. The road linking it to Halifax was based on an early Mi'kmaq route to Cobequid Bay from the Atlantic coast. First cut through in the 1760s, the "Old Cobequid Road" took many years to construct. In 1796 parts of it were said to be "nothing more than a blazed trail," but by 1816 it had become a stagecoach route. This route varied considerably in some areas from that of the present-day Trunk 2, although a few sections have remained relatively unchanged, and some parts of the now disused route can still be traced. Like the Windsor Road, it has been replaced by a modern highway (the 102) as the direct link between Halifax and Truro.

The road to Truro was less settled than the Windsor Road. The most densely populated section was around Gays River, where a combination of natural resources and early access by means of the Shubenacadie River had enticed settlement even before the road became established. Much of the remaining route passed through uncleared, undeveloped territory that gave little indication of the fertile farmlands discovered later in this region. Inns were established at intervals along the Truro Road to serve the travelling public, just as they were along the Windsor Road.

The other main routes in Nova Scotia developed in their turn as wheeled traffic increased rapidly and stagecoaches were soon offering service over longer and more varied routes. As the roads east of Pictou were improved, extensions of the coach service were developed to Pictou and Antigonish and eventually to Guysborough. There were also attempts to establish a more direct route to Guysborough. The main road through northern Nova Scotia to the border was obviously important, though travel to Amherst was often by boat across the Minas Basin to Parrsboro rather than overland all the way from Truro. A direct route initiated by Governor Parr from Halifax to Annapolis proved to be impractical, but sections of it were constructed and some have survived. The roads along the South Shore developed only slowly, and for a long time there was not a good direct road between Halifax and Chester. We shall look at selected stretches of these roads that served travellers in the first half of the nineteenth century.

Road Construction and Maintenance

The work of road building in Nova Scotia in the late eighteenth and early nineteenth centuries bore little resemblance to the blasting and bulldozing that accompanies modern highway construction. Roads either followed existing footpaths and portage trails or were cut through the forest, where surveyors with compass, chain, and axe had previously blazed the route on the trunks of trees. It was the road-builders' task to cut down the trees, remove the stumps, and haul away the smaller rocks in order to clear the right-of-way. Axes, picks, shovels, horses, and oxen were the chief construction tools. Large, immovable rocks were simply bypassed, creating an extra curve in the road.

The road from Halifax to Windsor was originally only eighteen feet wide, but as the province's communication system developed, all routes that were considered Post Roads were expected to be at least forty feet in width. Drainage was important. Work on the Windsor Road in 1779 required:

> *a trench to be dug on the upper side of the road, when necessary on both sides; [the surface] to be rounded in the middle, at the hollow places to be bedded with stones where they are to be had, and when they cannot be had, to be brushed and gravelled; all bad hills, where they are judged necessary, to have the tops taken off and carried into the hollow to make an easy ascent; the trees and brush to be cut down twenty feet wide on each side of the road from the centre; all short turns and bad hills to be altered where judged necessary.*

This was the ideal; not all sections of Nova Scotia's roads met these exacting standards. Upgrading the roads was a continuous process, and many minor alterations can be traced from surveyors' maps showing recommended improvements which would provide an easier gradient for carriages or avoid swampy areas.

Before the advent of asphalt paving, even the best roads were dirt roads—muddy in wet weather, dusty in summer, frozen or snow-covered in winter, and often impassable when the spring thaw or heavy rain turned the surface into a quagmire. Initially, when most traffic was on horseback, little attention was given to gradients, but with the increasing number of buggies and carriages in the early nineteenth century, means had to be found to make the roads safer for these vehicles. This was achieved in two ways: either the road was diverted around the side of the hill where the route was not so steep or material was hauled from the top of the hill and used to build up the surface of the road at the bottom as described above.

Nova Scotia has many lakes and rivers, some of which were used for transportation by the early settlers following the example of the Mi'kmaq; however, when the time came to build roads, rivers and brooks became obstacles. Although the valley cut by a river often proved a good route for a road, the river itself could form a barrier between a traveller and his destination. As a result, bridges, ferries, and fords became part of the developing road system. The bridges varied from fairly elaborate structures—some of them covered—to primitive log bridges thrown over small streams. Wide rivers were served by ferries, and where the water was shallow and the terrain was suitable, fords could be established.

Maintenance of the roads was initially provided by "statute labour" with funding from the legislature. A bill passed in 1761 required that each township appoint "fit and discreet persons" to be surveyors of highways, organizing settlers living beside the roads to contribute to their upkeep. Owners of carts were required to provide them, together with their horses or oxen, to work four eight-hour days each year. Those without carts had to provide six days' work to repair damage to the structure of roads and bridges in their neighbourhood. Snow clearing was also the responsibility of the roadside settlers, and since there were no snowploughs, they worked with wooden shovels and sometimes used their cattle to beat down a track through deep snow over which sleighs could pass.

From the second decade of the nineteenth century, serious attention was paid to upgrading Nova Scotia's roads. By 1813 the province had three surveyors general for its roads, one responsible for each of the Great Roads and one responsible for the remainder. District supervisors were appointed to purchase equipment, direct statute labour, and negotiate for (or expropriate if necessary) gravel and logs needed for the work in their own area. Money for new road building was allocated by the government to pay for surveyors' salaries, labour, materials, and compensation to landowners for property expropriated for road construction.

When Lord Dalhousie first came to Nova Scotia in 1816, he found that government patronage in the area of road construction had already become endemic in the province. He was initially inclined to view this with tolerance. "It is very natural," he wrote in his journal, "that a little self interest and private friendship should have a place in the distribution of money granted from the publick [*sic*] purse to the repair of roads and bridges already made, and to the opening of new roads thro' the wilderness and forest lands." Over time, he realized this system was inefficient, and he advocated reforms to the current practices. His recommendations fell on deaf ears. As for the roads themselves, he wrote: "The great roads are good, but the byeroads [*sic*] scarcely entitled to the name. The Acts of Statute Labour are not enforced, the grant of money much squandered from want of proper control or inspection, & what is repaired this year equally requires it the next."

The next lieutenant-governor also took an interest in the roads. Governor Kempt initiated a further series of improvements in the 1820s. The 1840s again saw significant efforts to upgrade the highways, and further changes were made to many main roads.

Early on, distances along the roads were often indicated at intervals by mile markers. Most markers on the first Great Roads have been lost, but at least two have survived. One of them now stands on the grounds of the Fultz House Museum in Lower Sackville, where the Windsor and Truro roads diverge. This is a piece of dressed granite, with the distances to the destinations carved on its face. The other is on the grounds of the Uniacke Estate, where the number twenty-seven is carved into a huge boulder beside the remains of the old road. Others were probably blazed on trees and have long since vanished.

This rock near Mount Uniacke along the Windsor Road was used as a mile marker and still bears the number twenty-seven.

Dwellers along dirt roads today may have some idea of the conditions that travellers endured even on the better roads in the early 1800s; however, with no graders, snowploughs, or other mechanical equipment, keeping the roads passable at all was a daunting task long ago. It is not surprising that many people still preferred to travel by water if possible. When public stagecoach services were established on the province's two Great Roads in 1816, the evolution of Nova Scotia's transportation system took a turn for the better.

Traffic on the Roads

Early in the nineteenth century, when neither public nor private transport was readily available, many people who needed to travel did so on foot. Pedestrians were common on the roads and frequently covered far greater distances than most people today would contemplate. Horses and their riders also formed part of the traffic, as did ox carts, horse-drawn wagons, gigs, buggies, and for the more affluent, closed private carriages. Carriages and horses could be rented from livery stables, which served a similar function to today's car rental companies, or a vehicle might be hired with a driver under private contract. Under normal conditions, a horse or a wheeled vehicle could make the journey between Halifax and Windsor comfortably within a day, arriving in time for dinner. The trip to Truro meant an overnight stop at one of the inns along the road.

In winter, wheeled vehicles were replaced by horse-drawn sleighs, and individuals sometimes covered considerable distances on snowshoes. Lord Dalhousie mentions the postman from New Brunswick making his way by dogsled, but this form of travel does not appear to have been common in Nova Scotia. It seems that under favourable conditions, horse-drawn sleighs could make as good time on the roads as wheeled traffic. Lord Dalhousie travelled the forty-five miles to Windsor by sleigh in February 1817 in six and one-half hours, including a break halfway to change horses and presumably have a meal. On another occasion in 1820, he recorded a non-stop outward journey by sleigh from Halifax to Mount Uniacke (twenty-seven miles) that took only two and one-half hours, with an even faster return journey. On still another occasion, when there had been a fresh fall of deep snow, the journey from Windsor to Halifax took eleven hours.

In 1829 Sir George Head, on leaving Halifax, "hired a sleigh to take [him] and [his] servant as far as Annapolis, a distance of one hundred thirty-two miles, for which [he] was to pay twenty pounds, or eighty dollars, including the expense of returning the horse and driver to Halifax." The journey took four days, and he then hired another sleigh to take him to Digby, an overnight trip.

Traffic on the Windsor Road was heavier than on the Truro Road. There was of course commercial traffic, and a wagon service had existed since 1798 to carry goods between Windsor and Halifax. People travelled from Windsor to the capital and vice versa for a variety of reasons. The population of the Windsor area was growing and included merchants and professionals who might have business in the capital, while those in Halifax had interests in Windsor. Many of the King's College students came from Halifax families. There was less military traffic in times of peace, but still the road provided a link between Fort Edward and the military installations in Halifax. The farmers who lived along the road often travelled it to sell their produce in town. For people from Halifax travelling to the Annapolis Valley, Digby, or Yarmouth, the road to Windsor was the first stage of their journey. As mentioned, Windsor was also a port from which ships carried passengers and freight originating in the capital to other settlements on the Bay of Fundy.

The journey between Halifax and Truro was longer and more tedious, and not undertaken lightly. Once the road was properly established, traffic increased and people living alongside it enjoyed better access to the capital and its amenities. Settlers in the Gays River area could then travel to Halifax by road, although the water route through Grand Lake was sometimes preferable. Truro, a transportation hub, was on the way to and from both Pictou and New Brunswick, so long-distance travellers also used the road. The town of Truro, as the administrative centre for the region, was a destination for those having business at its courthouse. In the fall, its agricultural show attracted visitors from as far away as Halifax.

Although the tourism industry was not the significant part of the economy it is today, there were people from away, mostly gentlemen with business in the region, who travelled out of curiosity as well as necessity. Residents also enjoyed leisure-time travels. Lord Dalhousie sometimes combined business with pleasure, and while inspecting the region's development, would stop to enjoy the scenery or indulge in shooting game. Family holidays were not the common events they are now, but travel for pleasure—usually over short distances!—was not unknown. Horses and gigs carried families to church, and no doubt also on visits to friends and relatives. Sleighing parties were fashionable for Halifax's better-off citizens, including groups of women like Lady Dalhousie and her friends, who would travel over the snow-covered road with jingling bells as far as Fultz's Inn on fine winter days. All of these people contributed to the traffic on the roads.

The first steps towards regular public transportation came in 1815, when Isaiah Smith was granted a licence to run a stagecoach service between Halifax and Windsor. The first coach left Halifax for Windsor in February of the following year and took nine hours for the journey, which originally cost six dollars. This initiative was followed in 1816 by Ezra Whitter's service to Truro, complemented by an extension to Pictou. These stagecoaches were essentially part of the postal service, replacing the mail carrier on horseback who still carried mail to other parts of the province. These coaches also carried paying passengers and their baggage.

In May 1817 the *Acadian Recorder* carried an advertisement for a "four-wheeled covered Carriage" serving Halifax and Windsor for the season. It was driven by John Smith and ran "once or twice a week as passengers or Packets may offer." The Windsor coach left from "Mr. Levi Wells's on Lower Water Street" in Halifax at 6:00 AM on Thursdays and from Windsor on Tuesdays. The fare was four dollars and fifty cents, and the rates for carriage of parcels and other goods were to be negotiated with Mr. Smith.

The same paper carried a similar notice of Ezra Witter's mail coach to Truro and Pictou. The extension to Pictou was run by Jacob Lynds. This weekly service left Halifax at 2:00 PM. The journey to Truro involved an overnight stop and took twenty-four hours. The fare was twelve dollars, of which half was paid in advance when reserving a seat. Baggage was limited to twelve pounds per passenger, and packages had to be prepaid. The entire journey to Pictou took two and one-half days.

From these humble beginnings, the Western Stage Coach Company and Eastern Stage Coach Company were formed in the late 1820s, providing service to Annapolis and Pictou, respectively. With larger and faster coaches and as the roads improved, better service was possible. It was not long before there were coach services from Halifax to Yarmouth at one end of the mainland and stretching as far as Guysborough at the other end. Public transport was not always of very high quality. In 1833 Lieutenant E. T. Coke described the vehicle in which he travelled from Annapolis to Windsor: "The coach [was] only a second-hand American one, with 'Western Mail, New York, and Hoboken,' upon its doors; neither was it waterproof, the canvas curtains hanging down in long shreds, and flapping to and fro with the wind." The horses were equally unsatisfactory, but the coachman was "perhaps one shade more professional than his counterpart in the States." Nevertheless, the coaching industry flourished until it was superseded by the railway in the second half of the century.

Settlement, Agriculture, and Industry

With a few distinctive exceptions, the settlers who lived along the roads in the early nineteenth century were modest people living modest lives. Many were new immigrants who had come to Nova Scotia from various places and for various reasons, simply wishing to possess land of their own, which was readily available to those prepared to work hard. The development of the roads provided opportunities for them to obtain land grants and become property owners. The land, at this stage, was undeveloped and the vegetation consisted chiefly of forest, which they first had to clear in order to obtain building materials and to start their crops. Their first homes were usually cabins built from the trees they felled. As time went on they were able to erect larger, though still simple, houses.

There were also more prosperous city dwellers who had invested in country properties along the roads with easy communication to Halifax; these estates were worked by their employees. The proprietors lived there in summer months and visited them from time to time in other seasons. The best-known of these estates was Attorney General Richard Uniacke's, whose home on the way to Windsor was described by Lord Dalhousie, fresh from Britain, as "the only Gentleman's seat on the road." Such estates, though, were in the minority.

Prosperous landowner or not, before any crops could be grown, trees had to be felled, and in many areas rocks had to be cleared away. Without the benefit of modern equipment, this was hard and time-consuming work. Saws, axes, and shovels were the main tools of the would-be settler, supplemented when possible with a horse or an ox and primitive implements. Lord Dalhousie described the preparation of land at Mount Uniacke in the summer of 1817:

> *Mr. Uniacke had this winter cut down some one hundred acres of wood; he was now burning them in some places; in others this had been done; the great logs dragged off, and he was planting potatoes in hills...or sowing oats with grass seed...; this last was the most clumsy & strange performance I ever saw, by a sort of harrow, a log cut in the woods with some iron spikes driven into it just to scratch the ground over the seed. A pair of bullocks drag[s] it among the stumps & roots, & when it catches at anything, a labourer going behind it lifts it up & rakes it free. This done, the ground is left for ten years to what little grass may grow, & while the stems and trunks of the large wood rot & become more easy to take out. As no manure is given, nor a ploughing practicable, I shall be hereafter very much astonished if either Potatoes or oats come to maturity upon it.*

Most of the roadside settlers were subsistence farmers, and by Lord Dalhousie's time many of them had cleared enough land to grow a quantity of crops, which enabled them to be largely self-sufficient. Some even had a little surplus produce to sell at a market in town. Many were still engaged in cutting down trees in order to expand their fields and pastures.

By then, parcels of land had already been sold by their original grantees to people who could continue to develop the farms without the trouble of initial clearing. His Lordship wrote in 1819 of a farmer living along the Truro Road who had come to Nova Scotia because of hard times in Scotland:

> *He has…brought his wife & young family, too young to be of any use to him, but he has brought also a little money & has purchased five hundred acres for £300. On the farm is a miserable house and a good barn. [Two hundred] acres have been formerly cleared, but are grown up in bush again. He will have little difficulty putting this much into crop, and make it a valuable farm.…If he prospers he will shew an excellent example in practical agriculture & [be] of great service in the present state of the Province.*

Many of these new settlers had little or no farming experience, particularly under conditions whereby they had to cut their fields out of dense forest before they could start. To compound their trials, in many places, particularly around Halifax, the soil was poor and rocky. A Scottish immigrant, John Young, took steps to encourage appropriate farming methods to help people make the most of their land. Starting in 1818, he wrote a series of letters in the *Acadian Recorder* under the name Agricola, outlining what he considered were good agricultural practices that would enhance the productivity of people's farms by measures such as crop rotation and manuring the soil. He successfully advocated the formation of local agricultural societies that would encourage farmers to adopt these methods.

Of course, farming was not the only occupation of dwellers beside the Great Roads; however, the need for self-sufficiency was evident, and most households would have had a cow, some chickens, maybe a pig or two, and at least a small patch of cultivated ground. The industries that sprang up along these roads were dependent on the resources available—raw materials and power—and on the settlers' needs. There were sawmills, gristmills, tanneries, and quarries. In the days before steam power was developed in the province, mills and other industries were powered by water from the many rivers and brooks in the region. Blacksmiths were an essential part of the economy. There were also military installations, educational establishments, and churches. Forerunners to today's tourism industry, there was already a network of inns.

Inns

At the beginning of the nineteenth century, the legislative assembly encouraged the establishment of inns for the benefit of travellers. In 1801 grants at a cost of twenty-five pounds were offered to settlers who would provide accommodation along the main roads. Six years later this cost increased to thirty pounds. The recipients were required to maintain their premises for at least six years, to clear five acres of land, and to erect a lodging house of at least sixteen feet by twenty-two feet to provide food and accommodation for travellers. This included rooms, meals, stabling, and perhaps also a change of horses. Stagecoaches required fresh horses every fourteen or fifteen miles in order to keep up a good speed, and individual travellers also needed food and stabling for their horses on a long journey. Some inns also served as post houses, where the mail to and from the area was distributed and collected.

William Moorsom, visiting Nova Scotia in 1830, observed: "The inns in towns such as Windsor, or Annapolis, are much the same as we find in the larger villages of England. The country inns are usually detached cottages, of which the owner having originally commenced as a farmer, and looking to that occupation as his chief resource, is a very different being from his accomplished prototype in England." Despite the less than professional services offered, Moorsom confessed to admiring the simplicity of these little inns.

As well as catering to travellers, the more sophisticated inns near Halifax also served as destinations for people going on a day's outing or as venues for social events. Places like the Rockingham Inn and Fultz's Inn served meals and provided ballrooms and entertainment for their patrons. Parties would travel by carriage in summer and by sleigh in winter to enjoy a day out of the city.

The inns on Nova Scotia's early roads, though primitive by today's standards, were vital to travel. At a time when a journey that now takes little more than an hour by car could take an entire day or longer, food, drink, and overnight accommodation were essential to travellers. When horses were a common means of transport, either as mounts or to pull carriages, provision for their care was as necessary as gas stations are to us today; stabling and fodder were among the services provided by the inns. Coaches stopped at certain inns to pick up and set down passengers, and even a brief pause provided welcome relief from the discomforts of early road travel. Long before tourism became popular, the hospitality industry was developing in Nova Scotia.

The early nineteenth century was a period of dramatic development and change as new settlements were established and the road system improved to accommodate a growing population. The following pages will take you through parts of Nova Scotia as they once were, paying particular attention to the Windsor and Truro roads. To glean as complete a story as we might, we shall look at the landscape, the buildings, the natural resources, the industries, and the dwellers beside the highways, as well as the changes that gradually took place at this pivotal stage in Nova Scotia's history.

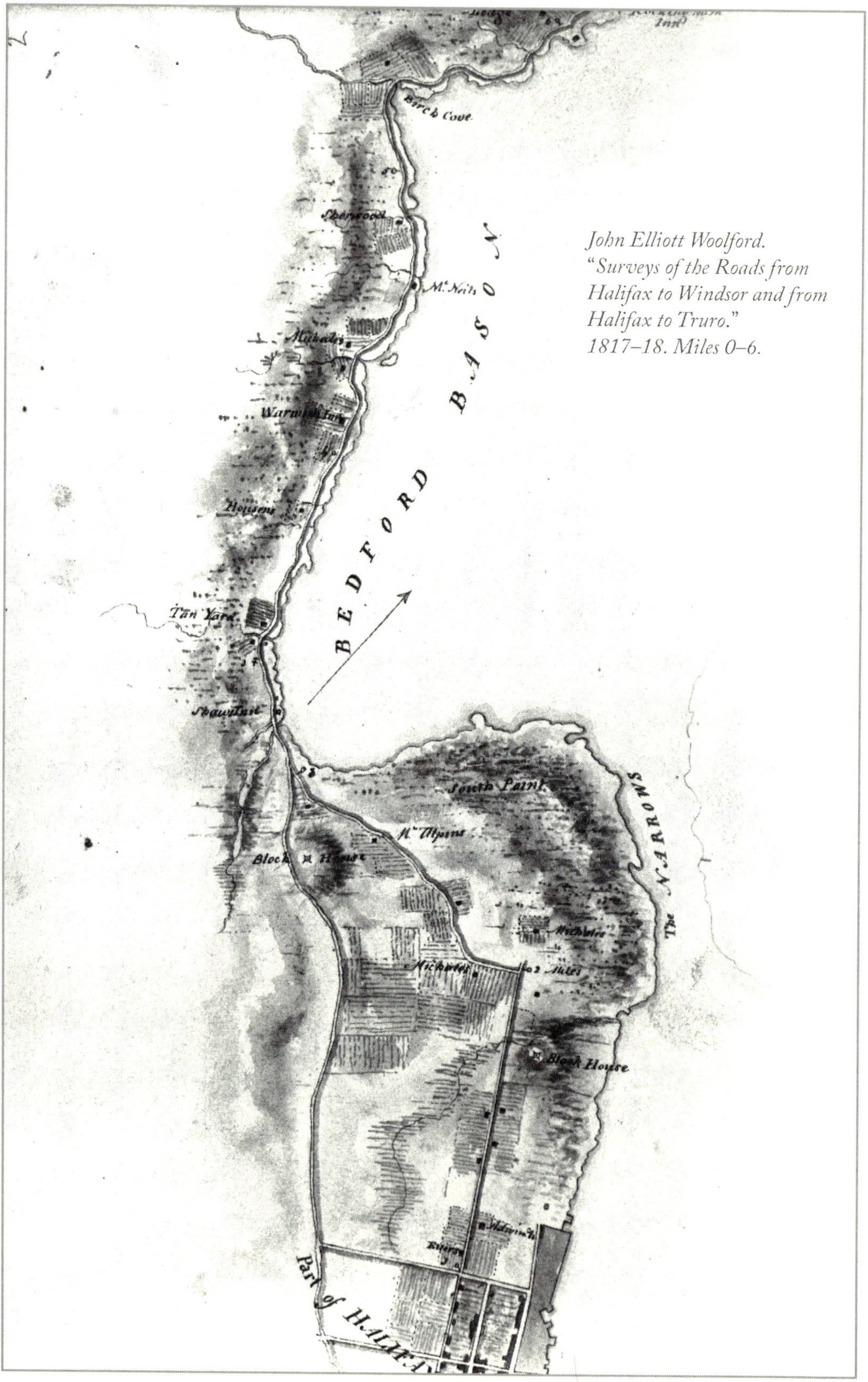

John Elliott Woolford.
"Surveys of the Roads from Halifax to Windsor and from Halifax to Truro."
1817–18. Miles 0–6.

1

The Road Out of Halifax
(Circa 1818)

Setting Out From the City

The major routes in nineteenth-century Nova Scotia radiated out from the capital as they do today. We shall follow the route travellers would have taken as they set out on their journeys. In those days, it seems people liked to get an early start. The coach for Windsor left downtown Halifax at 6:00 AM. Independent travellers also often started their journeys before breakfast, which they would take at one of the inns along the way. Only the Truro coach initially delayed its departure until the afternoon, since the journey would in any case necessitate an overnight stop. The time taken on the journey was very much dependent on the time of year and the state of the roads.

There was a choice of routes from downtown Halifax. The earliest was the Windsor Road, which corresponds approximately to what is now Windsor Street. It started from below the Citadel, crossed the Commons, and then ran much as it does today with a steep drop down to the Bedford Basin at Fairview Cove. Once past the Commons, it became a country road, running through land of which only some was under cultivation. The map indicates a number of cultivated areas, which by 1819 would also include John Young's thirty-acre farm, Willow Park. Young was notable for his letters published in the *Acadian Recorder* beginning in 1818, under the name Agricola (the Latin word for farmer). In these he expounded on his theories about agriculture. He established his farm in order to demonstrate what he deemed was good farming practice. His house and farm lay to the right of the Windsor Road, near the present site of the Halifax Forum. His influence is remembered today by the names Young Street, Agricola Street, and CFB Willow Park.

The other significant feature along this section of the road was the pentagonal blockhouse that also stood on the right, overlooking Fairview Cove. Known as McAlpine's Blockhouse, it was part of the early city defences that stretched across the peninsula to the head of the Northwest Arm. It had been constructed in 1808 on land owned by former soldier John McAlpine, who lived nearby. The blockhouse was built of heavy timber with an overhanging upper storey furnished with two ship's guns and barracks accommodation below. Its purpose was to protect the Windsor Road approach to Halifax.

After passing this structure, travellers of the day descended a steep slope down to the Bedford Basin, where the Windsor Road was joined by Lady Hammond Road (today's spelling) that came in from the right. This alternative, less direct route out of town was named for the wife of Lieutenant-Governor Sir Andrew Snape Hamond. It was built to link Gottingen Street with the Windsor Road in the early 1780s.

By the early nineteenth century, Gottingen Street provided the better route out of the city. This road ran from the foot of the Citadel almost due north through what was known as Dutch Town, where a number of the Foreign Protestants had settled in the 1750s. Only a few houses stood in the North Suburb beyond Gerrish Street, and much of the land on each side of the street was still under cultivation in about 1818. The first house identified on Woolford's map, at the corner of North and Gottingen streets on the left side, had been recently built by Benjamin Etter. It was just a mile from the Grand Parade, near where Mr. Etter owned a watchmaker and hardware shop at the corner of George and Barrington streets. After Etter's death, the house was sold to William A. Black, who was also a watchmaker.

The next section of Gottingen Street was dominated by elements of Halifax's naval and military infrastructure. On the right side, a little farther along, stood the stone-built Admiralty House that was completed during Lord Dalhousie's term of office. Admiralty House was begun in 1814 or 1815 as a residence for the British admiral commanding the North American station. It overlooked, as it does today, the Narrows and the Naval Dockyard. The house was transferred to the Canadian Government in 1904 with the rest of the dockyard and is currently home to the Maritime Command Museum. By the shore in the dockyard itself and visible from the road, stood the "shears" or shearlegs, a massive wooden structure in the careening yard that was used for tilting ships over in order to clean barnacles and other detritus from their wooden hulls. Beyond the masts of the moored naval vessels, the Town of Dartmouth could be seen on the far shore. A little farther north, the dockyard was protected by a blockhouse on the right side of Gottingen Street at Fort Needham, on the hill that still bears that name.

In Lord Dalhousie's day, Gottingen Street came to an abrupt end two miles from the centre of town, just beyond Fort Needham. The area north of the fort is where, in the days of Sir Andrew Snape Hamond, the "Governor's north farm" had served as a country retreat for the lieutenant-governor and his lady. (A house in this location is shown but not identified on Woolford's map.) To facilitate their access from here to the Windsor Road, Lady Hammond Road was constructed. It turned westwards between the farms of John and James Merkel (identified as Michale on the map) and sloped down towards the Windsor Road. The Merkel family's presence is commemorated today by Merkel Street and Merkel Place in Halifax's North End. This route to Fairview Cove was less precipitous than the Windsor Road and more suitable for horse-drawn vehicles.

On the left, near the bottom of Lady Hammond Road, stood the house of Captain John McAlpine, an interesting Scotsman who immigrated originally to upstate New York in 1773. After a series of misadventures during the American Revolution, he came to Shelburne with other Loyalist settlers in 1783. From there he moved to Liverpool, where he married a local widow, ran a tavern, operated a ferry, and among other occupations, was a deputy surveyor of the King's Woods. He became unpopular when, as deputy naval officer for Queen's County, he tried to enforce government regulations on the area's fishermen. After a bitter dispute with his neighbour and the death of his wife in 1790, McAlpine left Liverpool for Halifax and another phase of his life. During his first years there, he undertook various forms of employment to supplement his meagre military pension. He was charged with assault on several occasions and spent time in jail. By 1797 he was successfully running on his property a very fashionable inn known as Edward's Valley Inn, which gave his clients a magnificent view over the Bedford Basin. His house on Lady Hammond Road burned down in 1819, and he lived at the inn while it was rebuilt. He died in 1827.

The hilly area across the road from Captain McAlpine's property was identified on Woolford's map as South Point, south, presumably, of the Bedford Basin. On its shore, the beginnings of a little settlement of black refugees from slavery may already have been visible from the roadway, across Fairview Cove. Many of these people had come to Nova Scotia during the War of 1812. It is not known exactly when the first residents arrived in what would become Africville, but the town was officially recognized in 1840, after settlement had become well established. The community was demolished in the 1960s, and its residents were relocated under the guise of urban renewal. The area was recognized as a national historic site in 2002.

The Road Along the Bedford Basin

The Windsor Road and Lady Hammond Road converged as they do today, just three miles from central Halifax, to form a single road along the Bedford Basin. This road bore very little resemblance to today's busy Bedford Highway. Before the railway was built, the road ran closer to the shore, and travellers would have had an unimpeded view across the water. The road followed the shore to the head of the basin with wooded hills on the left, among which a scattering of dwellings—some inns, a little early industry, and a few large estates—had been established.

Just beyond the intersection of the Windsor and Lady Hammond roads was Shaw's Inn or Three Mile House. It had been built by a Mr. Shaw, a member of the legislature, and it marked the beginning of the journey along the Bedford Basin. Today's busy highway looks nothing like the quiet thoroughfare in this sketch by John Elliot Woolford depicting a horseman riding along the rutted surface and two pedestrians making their way safely down the middle of the dirt road.

The next feature on the road was the tannery that stood to the left. The "tan yard," as it was called, was run by a man named Stanford and is the first evidence of industry developing along the road. It stood beside a brook that ran into the Bedford Basin and was powered by water from the upper section of the brook. The water was carried by a flume from which it poured onto the overshot wheel. The tannery itself was an extensive building, and the house beside it, with its neat walled gardens, was presumably the owner's residence. The building on the water side of the road appears to have been a barn or stable.

There were several residences along this section of the road, some modest, others spacious estates. The Housen family lived on a small property on the left side, between the tannery and Warwick Inn, which was otherwise known as the Four Mile House. The inn stood on a large cultivated lot, which would have provided some of the produce needed to serve its clients. The next property is identified on the map as "Michales," but as previously noted, this spelling seems to refer to the Merkel family. Their house stood among cultivated fields on a small stream. A little farther along, the McNeil home also stood by a stream on the shore side of the road, with no apparent cultivation.

The next section of the road featured several impressive country properties established by well-to-do Haligonians. Next on the left, just before mile five, was the large estate known as Sherwood. The house stood in a clearing on the wooded slope above the road. The house is gone, but its name endures in the Sherwood Park subdivision. Originally built for Bishop Stanser, the house was at this time owned by Mrs. Jane Donaldson, the prosperous widow of Thomas Donaldson, a confectioner who had established a fashionable store on Granville Street. After her husband's death, Jane Donaldson continued to operate their store.

According to Dr. Akins's *History of Halifax City*, Jane Donaldson owned the Sherwood estate while living a little further along the road, in the house at Birch Cove that Lieutenant-Governor Sir John Coape Sherbrooke had previously

John Elliott Woolford. "Road Scene on the Bedford Basin near Shaw's Inn." (Nova Scotia Museum 78.45.20) The building in the centre of the photo is Stanford's tannery.

Woolford, "Surveys," miles 6–12.

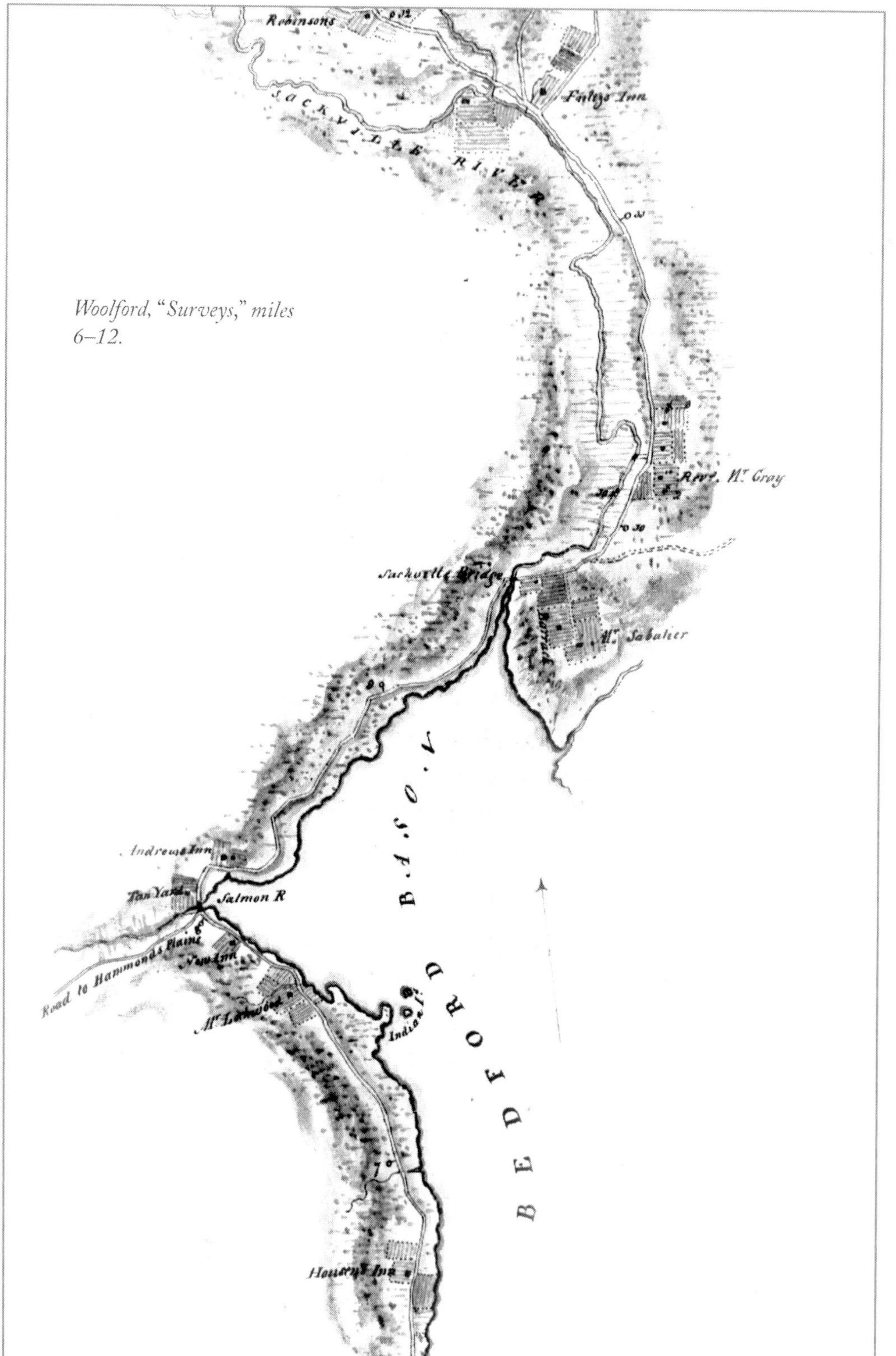

lived in. The Birch Cove property also stood on extensive grounds, at the mouth of the brook that ran into the cove. It remained in the Donaldson family for some time. Today's Dakin Drive traces the course of a portion of the old road that ran along the cove.

Six miles from the centre of Halifax stood the Lodge, formerly the residence of Prince Edward, Duke of Kent, during his time in Halifax when he served as commander-in-chief of British troops in North America. The property belonged to the lieutenant-governor of Nova Scotia, Sir John Wentworth, and had been developed by the prince soon after his arrival in 1794. Under his direction, the house and grounds took shape with carriageways, walks, and summer houses set among the trees. A heart-shaped pond was created for the delight of his mistress, Julie de Saint-Laurent, and a music room in the form of a rotunda overlooked the basin. The prince also had buildings erected nearby to house some of his troops. On his departure for England in 1800, the property reverted to Wentworth, who continued to occupy the Lodge for much of his remaining life. After his death in 1820, it fell into disrepair, but the music room on its hill to the right of the road remains a landmark. Today's visitors still enjoy Julie's Pond, situated at the entrance to Hemlock Ravine Park.

The Rockingham Inn, just visible at the top of the map and standing between the road and the shore, occupied one of the former military buildings that Prince Edward had constructed. The inn was close enough to Halifax to be a fashionable destination for a day's outing, equipped with a dining room and a ballroom, as well as providing overnight lodging. Its accommodation was of sufficiently high quality that Lord and Lady Dalhousie spent a week there in October 1818, while some renovations were carried out at Government House. In Wentworth's day it was known as Rockingham House and became the venue for a literary and social club, which he established for Halifax's élite. By Lord Dalhousie's time the club was in decline, but in 1818 His Lordship revived it with a dinner for about thirty members who, he said, spent a very pleasant day there. Although the lodge and the inn are gone, the names Prince's Lodge and Rockingham are both preserved in nearby communities.

Then as now, the section of road north of Prince's Lodge provided ample sites for refreshment of passersby. There were three more inns between the Rockingham Inn and the head of Bedford Basin: Housen's Inn was a little before mile seven, and New Inn and Andrew's Inn were a mile farther along with one on each side of the mouth of the Salmon River (now Mill Cove).

On the left side of the road past Housen's Inn, a promontory could be seen beyond which lay two small islands; one of these is identified on Woolford's map as Indian Island. They are less visible today from the highway but can be seen clearly from the train. Just beyond this point, on the left, stood another large property, identified as belonging to "Mr. Lockwood," who seems to have been an unusual character. As a young man, Anthony Lockwood had been a British naval officer and a hydrographer, contributing to Admiralty navigation charts of British North America. Upon settling in Nova Scotia sometime before 1817, he had worked as an assistant to Charles Morris III, the surveyor general. He also established a private surveying business, advertised in 1817 under the name "Lockwood and Son, Land Surveyors and Draughtsmen," although his son, also named Anthony, was barely into his teens at this time. The following year, Lockwood had a book published at his own expense in London, entitled *A Brief Description of Nova Scotia with Plates of the Principal Harbours*, dedicated to Lord Dalhousie and to the surveyor general. He identified himself on the title page as "Professor of Hydrography and Assistant Surveyor General of Nova Scotia," though these formal titles may have been exaggerated. In October 1817 he advertised for sale both his house and some surveying instruments and stated his intention to leave the province. But he still owned the house when Woolford was preparing his map later in 1817, and he was again advertising his surveying services the following spring. The house was once more for sale, along with some mathematical instruments, in August 1818. In 1819 he published in Halifax a map of Nova Scotia, printed by a local engraver, C. W. Torbett. He presumably retained his surveying instruments, too, as in 1819 he sought and received the appointment as surveyor general of New Brunswick. Perhaps this vacillation was an indication of the mental problems from which he would later suffer. He took up his duties in New Brunswick in July of that year, so his house would certainly have been vacated before Lord Dalhousie's term of office came to an end in 1820. Once in New Brunswick, Lockwood alternated periods of great productivity with periods of madness. He eventually returned to England, where he died in an insane asylum.

The next landmark on the road was the inlet where the road crossed the Salmon River, scarcely recognizable today because of the vast developments around Mill Cove and changes to the highway. The commercial potential of this area was exploited even in Lord Dalhousie's day. In addition to the two inns already mentioned, this was also the site of Mixner's Tan Yard. This industry, like the tannery at Fairview Cove, was powered by water, in this case from the Salmon River. The two inns were strategically located at the junction of the Bedford Road with the Hammonds Plains Road. In those days, the road ran to the south of the Salmon River.

The Hammonds Plains Road was for many years a major route from Halifax to the South Shore. It led towards the head of St. Margarets Bay, as it does today, and was originally intended to branch cross-country to Annapolis. Because of the inhospitable terrain, only small sections of the Annapolis Road were completed. The Hammonds Plains Road continued as a link to what became Trunk 3 and now also to Highway 103. Refugee slaves from the Potomac and Chesapeake rivers who had come to Nova Scotia in 1814 were establishing themselves on government land near the road at Hammonds Plains (named for the former governor), which soon grew into a predominantly black community. By 1819, a number of the black settlers had built houses and were petitioning for title to their land. Today Hammonds Plains is a racially diverse community and part of Halifax's urban sprawl.

Andrew's Inn, just north of the tannery, marked the end of development along the Bedford Road before it reached the bridge over the Sackville River. On the far side of the river was the area where Fort Sackville had been established soon after the founding of Halifax. Its buildings remained there in Lord Dalhousie's time, to the right of the road, marked on Woolford's map simply as "Barracks." The fort consisted of a fenced area with barracks and a storehouse and a guardhouse which served less to protect Halifax from attack than to prevent would-be deserters from leaving town by this route. The site can still be identified, but there is little to see today.

Property owned by William Sabatier, whose name appears on the map, lay just to the east of the military ground and perhaps included some of its buildings. Sabatier's house had been built by Joseph Scott, an early settler, saw- and gristmill owner, militia officer, and for some years a Justice of the Peace, Justice of the Inferior Court of Common Pleas, and holder of various other offices in Halifax, including that of "surveyor of lumber." Scott first came to Halifax in 1749 and established a hardware store, which soon expanded to carry lumber. He acquired a huge tract of land in the Sackville area from which he cut timber to supply his sawmills on the Sackville River, which in turn supplied his hardware business. At some point in the 1760s, he sold his Halifax home to move to his property at the head of the basin, where he built the "manor house." He spent the rest of his life there.

William Sabatier, a Halifax merchant, first leased the property from Scott's widow and later purchased it. He initially farmed the land, but by 1814 he had already moved back to Halifax, where he served for a while as sheriff. He returned to England in 1819, leaving the sale of the property to one of his business associates. The house, a fine gambrel-roof building on pleasant grounds, changed ownership several times but remained a private residence until the late twentieth century, when it became a community museum known as Scott Manor House.

Ten Mile House to Fultz's Inn

The inn known as the Ten Mile House stood nearby, though it is not identified among several buildings shown on the map as part of the manor estate. Thought to date back to the 1790s, the inn may have formed part of the early barracks complex. This building stood beside the road just beyond Fort Sackville and became one of the many inns serving the travelling public. Like other inns near Halifax, it was also a destination for people seeking a day out of town. The building has passed through several hands since the early nineteenth century. At one point it was occupied by Ambrose Church, publisher of Nova Scotia county maps in the latter part of that century. It is still extant but had to be moved back when the road was widened, so its well was left at the front of the building. At one time an art gallery, later a music shop, and now a cycle shop, it has been somewhat defaced by commercial signage.

The Ten Mile House, 1980s.

Just beyond the Sabatier estate, a trail or minor road can be seen on Woolford's map branching off to the right. Comparison with contemporary maps shows that it led in the direction of Rocky Lake. Now unrecognizable after considerable modifications, it is known today as Rocky Lake Drive and has replaced the Old Cobequid Road as the initial section of Route 2 to Truro.

Well at Ten Mile House, 1980s.

Following the Sackville River, the Windsor road took travellers past the Parsonage House, home of the Reverend Benjamin Gerrish Gray. He was the first rector of St. John the Evangelist Anglican Church in Sackville. His house stood high above the road on the right, the first of three such houses shown on the map. Gray and his neighbours had cleared land around their homes and probably used the meadows on both sides of the Sackville River to graze their cows. A small road led down to the river with a bridge that gave access to the far side.

Gray was born in Boston and came to Nova Scotia at the age of eight with his parents. He was ordained in 1796 and appointed chaplain to the Maroons in Preston by Governor Wentworth before he became a teacher and acting librarian at the academy of King's College (later Kings' Collegiate School at Windsor). He began serving the Sackville parish under the auspices of the Society for the Propagation of the Gospel (SPG) in 1806. Although he had acquired considerable property in the Sackville area, he would soon move to St. George's Church in Halifax. His final appointment was as rector of Trinity Church, Saint John, New Brunswick, where he died in 1854.

After this small group of houses, the road passed through uninhabited countryside for about a mile before reaching Fultz's Inn. The inn, also known as Fultz's Tavern or the Twelve Mile House, had been established in 1814 by William Fultz whose ancestors were among the early settlers Cornwallis brought to Nova Scotia. Gray had sold a parcel of land in this area to William's father, Anthony, including the lot on which the inn stood. Other members of the Fultz family would settle there over the years, giving the name Fultz's Corner to the intersection that marked the divergence of the Windsor and Truro roads.

The inn served not only travellers to and from both destinations but also Haligonians enjoying a day in the country. The establishment of stagecoach traffic to both Windsor and Truro in 1816 increased the number of visitors to the inn. It became a fashionable summer resort and, like similar businesses of its time, offered a large ballroom as well as food and lodging. It was also a popular destination for sleighing parties whose jingling bells could be heard on fine winter days. Travellers, including Lord Dalhousie and his companions, frequently stopped at Fultz's Inn for breakfast after an early start on their way out of Halifax. Its reputation had not diminished by 1829, when Joseph Howe observed that it was "not so much a matter of course to get well fed at Fultz's, as to have a good appetite when you get there."

In November 1816, shortly after his arrival in Halifax, Lord Dalhousie rode out to breakfast at the inn. He was struck by the devastation caused by recent forest fires in the area. "We walked out after breakfast for a couple of hours to look at the awful ruin of the woods by fire and strong winds. All the large wood

M. Seymour, Fultz's Inn, 12 Miles from Halifax.

is laying black and tore up by the roots, in masses of confusion. Others remain erect but dead and rotting, to fall by the next blast. There is however some fields cultivated." Fires continued to break out along the Great Roads over the following years, and their ruinous effects were noted by Lord Dalhousie as he recorded his travels.

The milestone from the intersection of the Windsor and Truro roads is now on the grounds of the Fultz House Museum. J. Dawson photo.

Although Fultz's Inn was frequented by Joseph Howe in later years, the coming of the railway, for which Howe was in no small measure responsible, reduced road travel. During the second half of the nineteenth century, the house ceased to function as an inn and became a private residence. It burned down in 1890. The original structure stood at the far side of the intersection, across the road from the present site of the Fultz House Museum, where many relics have been preserved. Among these is the mile marker that in Lord Dalhousie's day stood at the intersection of the Windsor and Truro roads, near the Robinson property. This stone marked the end of the first section of the road out of Halifax. From here travellers to Windsor and to Truro went their separate ways and encountered very different features along the roads.

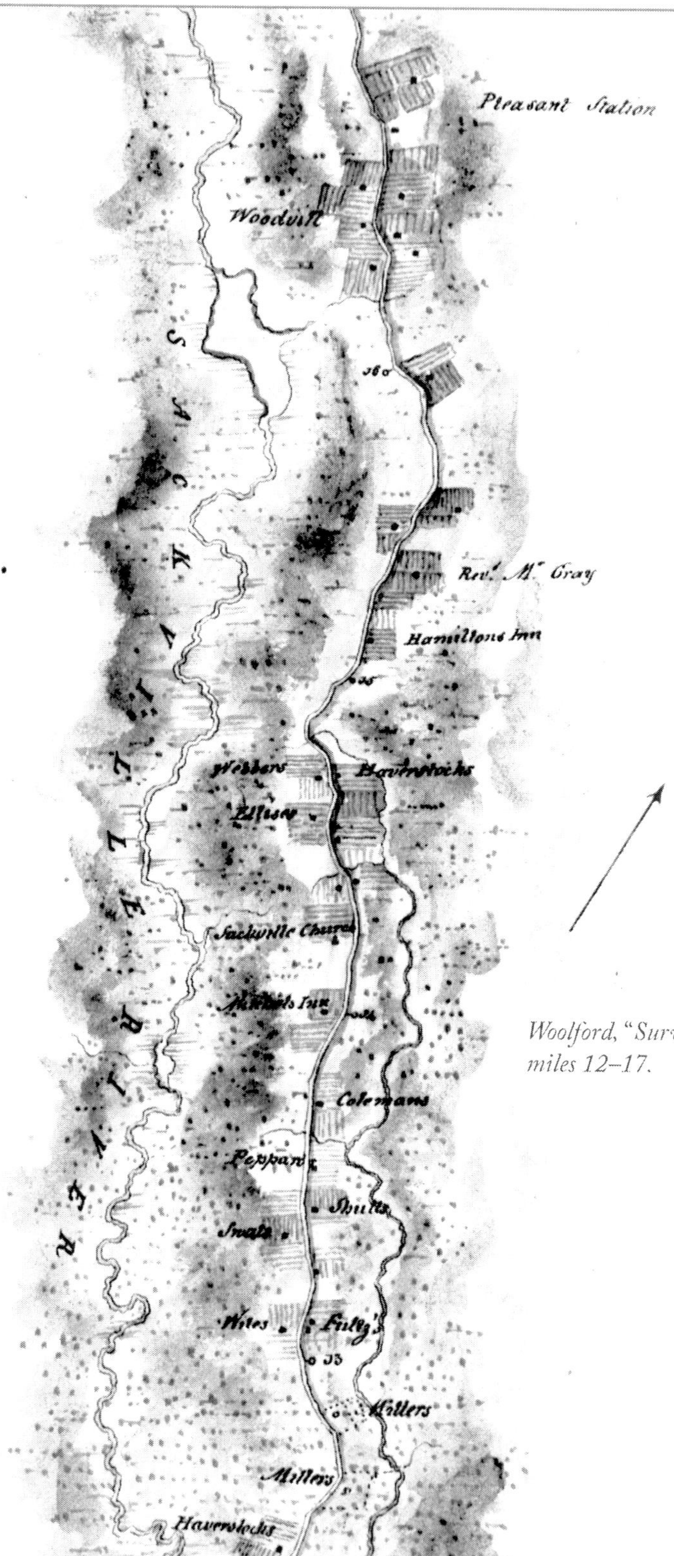

Woolford, "Surveys," miles 12–17.

2

The Road to Windsor
(Circa 1818)

Leaving Fultz's Inn, the early nineteenth-century traveller to Windsor would bear left while the Cobequid Road to Truro branched right. The Windsor Road originally followed the ridge between the Sackville and Little Sackville rivers, avoiding the marshy riverbanks. Known today as the Old Sackville Road, this section was superseded as the main road to Windsor by a lower road running parallel to it. Trunk 1 still follows the general line of this lower road, but much of its remaining route to Windsor is very similar to the road represented by Woolford's map. Trunk 1, in turn, has been replaced by Highway 101 as the most heavily travelled route between Halifax and Windsor.

Lord Dalhousie described the original road to Windsor as he first saw it in February 1817, after the previous years' ravages of forest fires and storms:

> *From Halifax to the top of Ardoise Lake 30 miles, the soil is rocky & extremely bad in general—a great quantity of wood, but from wild fires & sweeping hurricanes of wind, half of it is erect but burnt, while the rest is blown down, uprooted, being now decayed. The habitations, however—little inns and cottages—are very thickly set.... The moment we begin to descend towards Windsor, the country assumes a new appearance. The Ardoise is a ridge of high land rather than a hill, extending a considerable*

distance East & West. On the Windsor side no fires appear; a great number of fine trees are laying blown down or leaning over dead...which gives a wildness to the scene very striking. The woods are perfectly thick & impenetrable, unless to an Indian—mixture of spruce, fir & hemlock with hardwood of Oak, Birch, Beech and maple, very lofty but not large dimension, from being so thickly set. Some few larch, here called Hacmetac [sic]. Numbers of very small cottages with a few acres cleared around them, and another piece in process of clearing, shews the manner in which this Province is increasing, both in the number of Cultivators and in quantities of arable land.

The Road through Sackville to Mount Uniacke

By Lord Dalhousie's time, settlers were already well established along the first section of the road. From the properties owned by the Robinson family near the confluence of the Sackville and Little Sackville rivers (at the top of the previous map) to the cluster of houses at Woodvill and Pleasant Station, the road served as Sackville's main street. For about three miles there were simple houses on both sides of the road, each with a small area of cleared and cultivated land along the ridge between the two rivers. The population at this time consisted chiefly of subsistence farmers whose surplus produce supplied the Halifax market.

In the 1820s, in order to avoid the hills, the road that ran along the valley of the Little Sackville River was developed as the main highway to Windsor. It bypassed the former centre of the settlement, as Trunk 1 does today. In 1828 Joseph Howe wrote in his *Western Rambles*: "The first thing that strikes the eye after leaving Fultz's is the alteration of the road, by which so many hills are avoided, and many houses left in the rear; and when you think of the pleasant view you might have had from the top of the hill, you are half inclined to regret the alteration—but so it is, utility and uniformity will triumph over nature, and the picturesque must suffer from the change."

The property owners of Sackville were mostly hard-working, undistinguished people who have left no major mark on history, though all the names that appear on the map can still be found among the area's population in the early twenty-first century. Some of their houses survived into the later nineteenth and early twentieth centuries, but most have long since vanished. Robert Harvey has traced the history of some of these houses and their owners in his work, *Historic Sackville*.

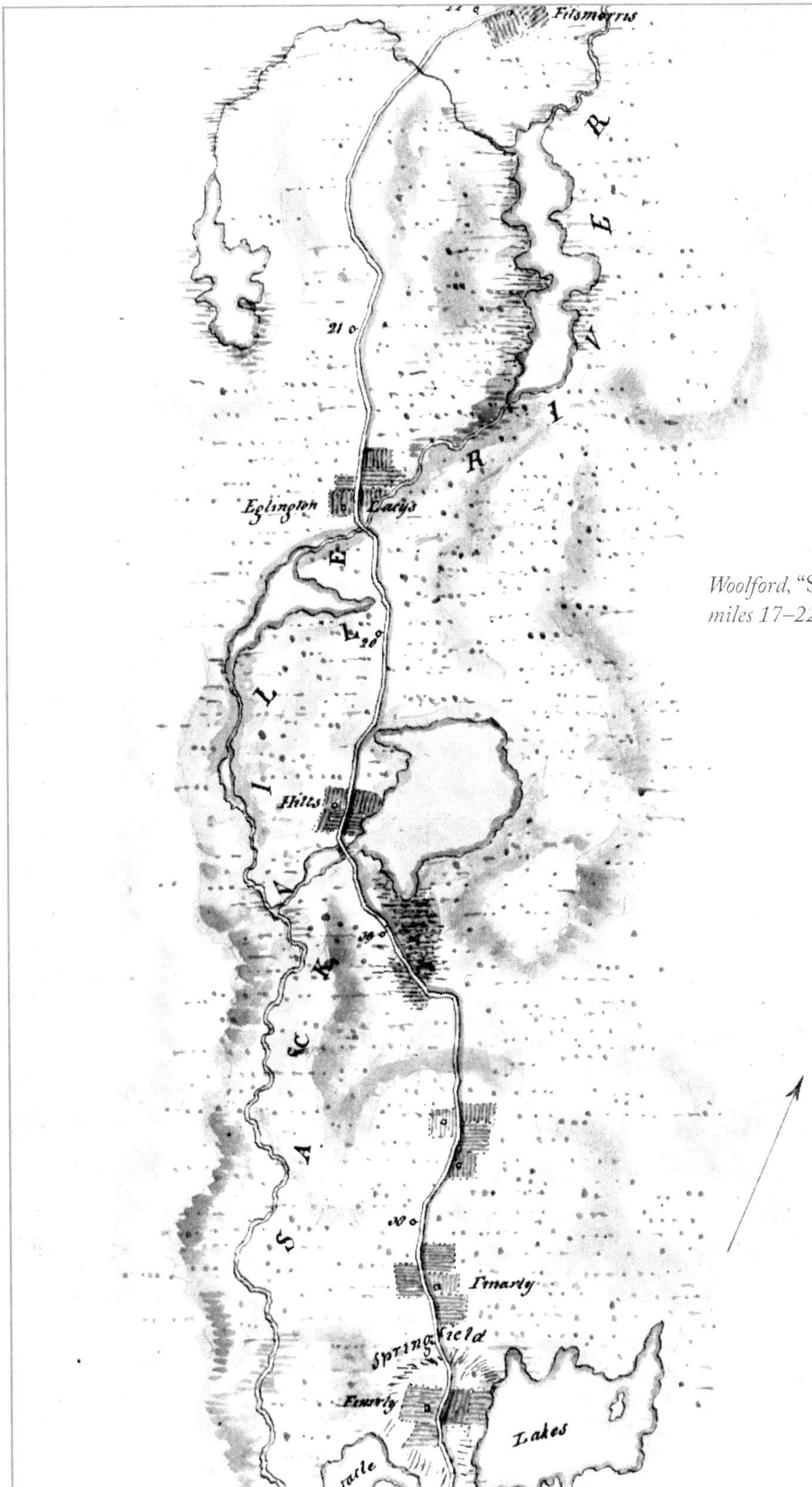

Woolford, "Surveys," *miles 17–22.*

As well as modest farmhouses, Sackville's buildings included two inns and a church. Mitchell's Inn at mile fourteen (the distance at which stagecoaches usually changed horses) and Hamilton's Inn a little farther along the road were the two places for entertainment in the area. At the centre of the settlement, just beyond Mitchell's Inn, the Sackville church offered the nearest approach to a focal point for the community at that time. Built in 1806, St. John's Anglican Church and its graveyard stood on an elevation to the left of the road. This early building, where the Rev. Mr. Gray conducted services, burned down in 1828 and was replaced by the present-day church. As previously mentioned, Gray had invested in land in the Sackville area; in addition to the parsonage, he owned a farm just beyond Hamilton's Inn to the right of the road.

The road continued with small farms on either side, past a small lake on the left we know today as Webber Lake and through the cluster of houses and fields identified at the top section of the map on page 32 as Woodvill and Pleasant Station. The original road and Trunk 1 come together again at Upper Sackville. This whole area has since been absorbed into the larger communities of Upper, Middle, and Lower Sackville.

The road then passed between the Spectacle Lakes (today Springfield and Drain lakes) to the property known as Springfield. Here members of the Fenerty family had established themselves with a homestead and several fields under cultivation on both sides of the road on a hill overlooking the lakes, along with a second house and fields at the foot of the hill, just before mile eighteen. The original grant to William Fenerty dated back to 1784, and his heirs added to it. William and his descendants successfully farmed, cut lumber, and ran sawmills on the nearby lakes. In Lord Dalhousie's time, Springfield also served as an inn, otherwise known as Fenerty's.

The most famous Fenerty was William's grandson, Charles, born in 1821, only a few years after the map was made. Charles was the son of James and Elizabeth Fenerty, who occupied the homestead at Springfield. He is remembered for his discovery that paper could be manufactured from wood pulp, thus contributing to the development of the newsprint industry. He is said to have watched wasps gathering the "fuzzy material" created by friction of the heavy wooden machinery at the family sawmills to construct their nests, and he realized that wood could be substituted for the hard-to-obtain linen rags used in the manufacture of paper. Also a writer and a poet, Charles travelled to Australia before returning to the Sackville area to marry and settle down to life as a farmer.

Between miles eighteen and nineteen, the woods on each side gave way to a swampy area beside a small lake (Lewis Lake today) and shortly afterwards, the road crossed over an arm of the lake that fed into the Sackville River. The main road has been straightened in this area, but the older road still exists as a loop to the right just after Springfield Lake. Near Lewis Lake was the Hiltz's farm, with the house on the left side of the road and some of the fields dropping down to the lake on the right. A mile farther on, another bridge crossed the Sackville River, bringing the traveller to Eglington, an inn otherwise known as the Halfway House (between Halifax and Windsor), run by the Laceys. This was a well-known stopping point for travellers on the Windsor Road, just over twenty miles from Halifax and only slightly farther from Windsor.

Lord Dalhousie's first excursion to Windsor was made by sleigh in February 1817, and he recorded stopping at this house to change horses. "The roads," he wrote of the outward journey, "were well beat and smooth as ice the whole way." On the return journey, however, the road was clogged with snow. He reached the Laceys' and changed horses there only after the sheriff of Hants County had brought out three pairs of his own horses to beat down a path on the road as far as Mount Uniacke. He mentioned changing horses there again in August, this time in a gig, so Mr. Lacey's service was presumably satisfactory. The Lacey family has left its mark on the modern map at Lacey Lake, to the left of Highway 101.

Today, just beyond the lake, another turnoff to the right from the present-day road represents the original route of the Windsor Road from about mile twenty-one to mile twenty-four. The population here in about 1817 was rather scattered. A mile beyond the Fitzmorris farm, another inn—run by John Lunn and his wife Sarah—stood on the right side of the road, with McDonald's farm close by. It is another two miles to Pence's (Pentz's) Farm. Sometimes known as Pentz's Tavern, this was one of the oldest inns on the road and had been run by Jacob Pentz as a "house of refreshment" since 1802. Tenants of the time would see a moment of drama in 1819, when Richard John Uniacke, Jr., took a room there on the day of the duel in which he killed William Bowie. Family tradition had it that after he had left, the landlady, Ann Pentz, found a folded, bloodstained handkerchief with a hole through it.

Another mile through uncleared forest brought the traveller to the boundary of the large estate known as Mount Uniacke. Its owner, Richard John Uniacke, served as Attorney General of Nova Scotia from 1797 to 1830. In 1817 the house had been only recently completed and was described by Lord Dalhousie in

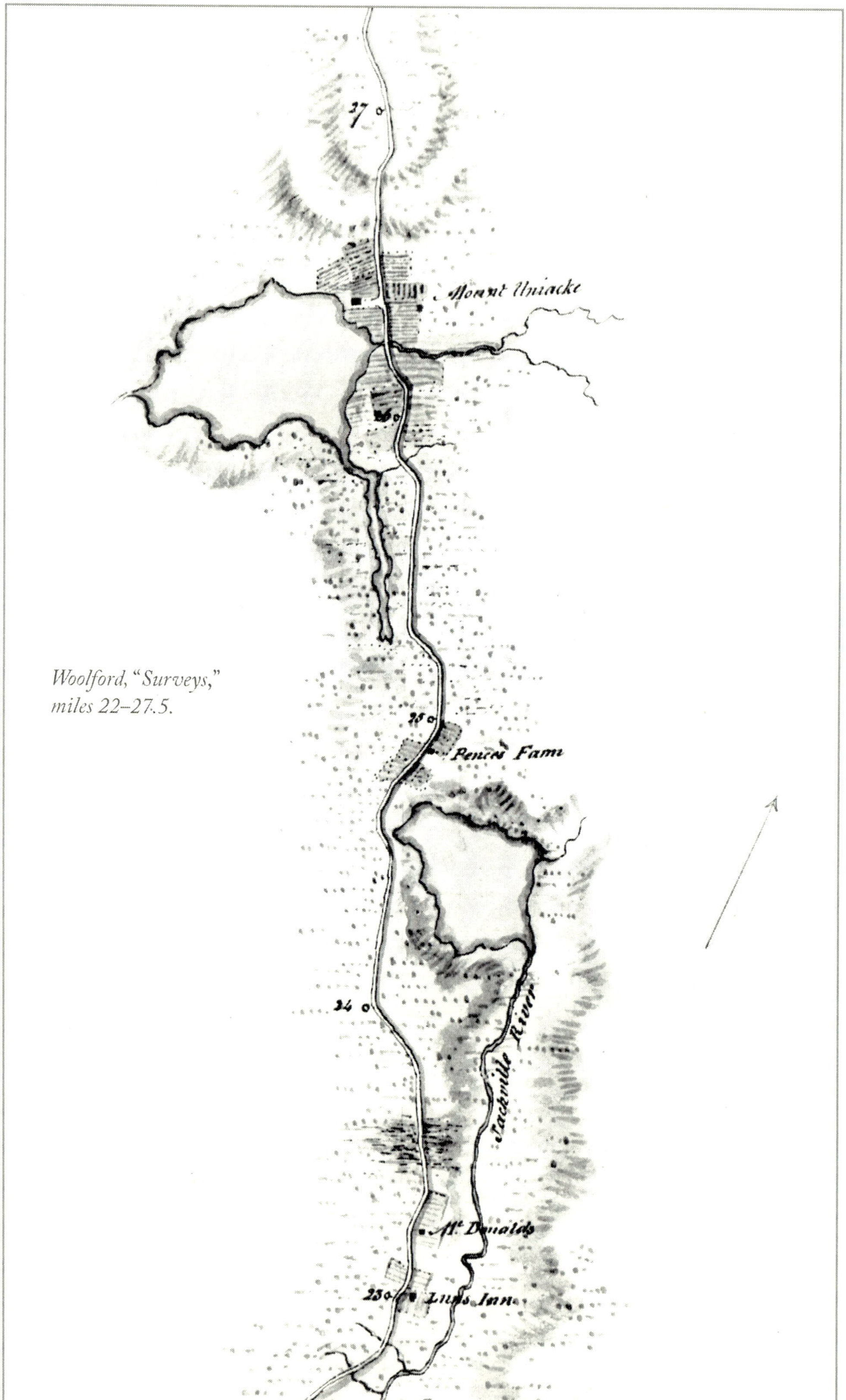

Woolford, "Surveys," miles 22–27.5.

that year: "Mount Uniacke is the only Gentleman's seat [i.e., country mansion] on the road, finished last year, has a very comfortable neat appearance but tho' called is not a Mount, but situated in a bottom between two very pretty and extensive lakes, surrounded with wood, Pine and hard." This was the Attorney General's country home, where he spent the summer and closed it up completely in winter, to His Lordship's surprise, "without even a Servant to keep a fire in it."

The house and grounds were an impressive sight from the road in 1817. An imposing building of neoclassical design with a pillared portico, it stood (and still stands) on extensive grounds with a coach house and other outbuildings. It sat back from the road, separated from it by a stone wall, and its garden side faced Lake Martha, named after Uniacke's late wife. As well as the garden, a large area around the house was cleared or in the process of being cleared for agriculture. Stumps still stood in one of the fields visible from the road, while animals grazed in an adjacent pasture.

Lord Dalhousie was a frequent visitor at Uniacke House. He appreciated the Attorney General's efforts to build a civilized residence in the middle of nowhere

John Elliott Woolford. "Mount Uniacke." (Nova Scotia Museum, 78.45.59) "The only Gentleman's seat on the road," according to Lord Dalhousie. The steep hill on the Windsor side, from which this sketch was made, was later replaced by a road with a gentler gradient.

but also observed that much still needed to be done. "Mount Uniacke," he wrote in 1817, "situated on the margin of a fine lake & surrounded by the woody wilderness mixed up with great granite rocks, is a very gentlemanlike, & may in time be a pretty, place, but at present has little to recommend it, except the new comfortable house and the cordial hospitality of its Proprietor." The following year His Lordship described his host as "a most pleasant, well-informed and sensible man. He lives a great deal at this seat of his own making, he is a zealous settler & spends a great deal of money on this place, which is pretty but thought to be sadly unprofitable to his family."

Unlike most of the buildings that lay along the Windsor Road at that time, Mount Uniacke still stands today, but the realignment of the road makes it invisible to passersby. The house and surrounding parkland trails are managed as part of the Nova Scotia Museum. The early route of the Windsor Road passed through what is today the Uniacke Estate Museum Park. At least two changes have been made so that the present Route 1 bypasses the estate, leaving the former road as a driveway and a walking trail, where part of the old stone wall is visible today. Near the trail, on a massive rock that once served as a mile marker, the number twenty-seven can still be read.

Mile Twenty-seven to St. Croix

After Uniacke House, the Windsor Road passed through forest again for another two miles before running between the two lakes that give the name Lakelands to the area today. Several houses stood on plots cleared and cultivated by the original settlers and watered by two streams. The property owners included the Lawrences, the Thompsons, who owned an inn surrounded by a large acreage of arable land, and a Mr. Major, who was clearly of higher social standing, judging from the honorific "Mr." This may have referred to Frederick Major, a Halifax businessman at the time, along with his son of the same name. The other residents of the area are not identified on the map, but Lord Dalhousie recorded that Thomas Jeffreys, the Collector of Customs, also owned property here.

The next mile or so of the road again passed through uncleared woodland and opened to a view on the left, over the lake known today as Cameron's Lake. (In the early nineteenth century it was called Hutchinson's Lake.) The cleared land on each side of the road here was the newly acquired property of Charles Morris III, the surveyor general of the province. When Lord Dalhousie first came to Nova

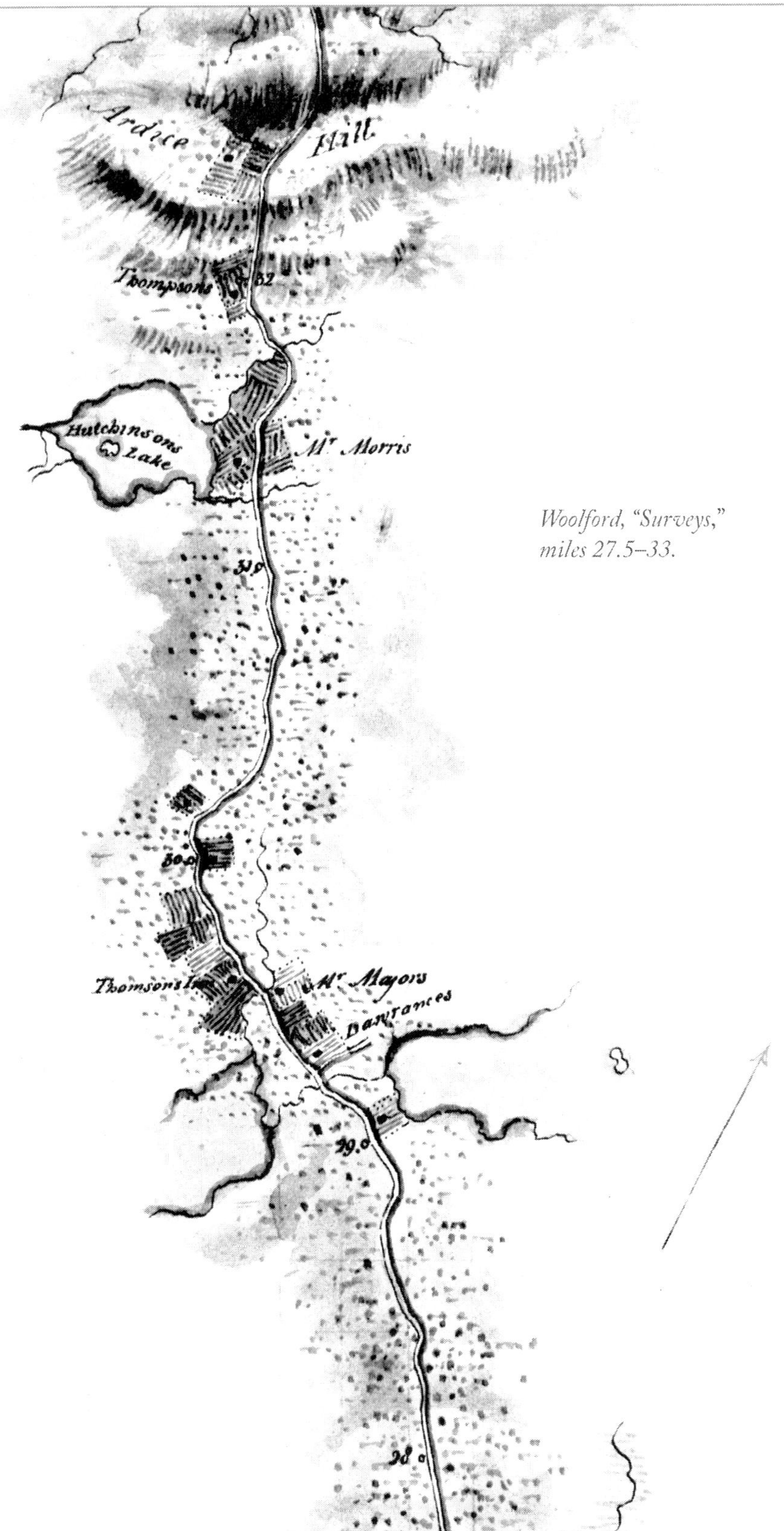

Woolford, "Surveys," miles 27.5–33.

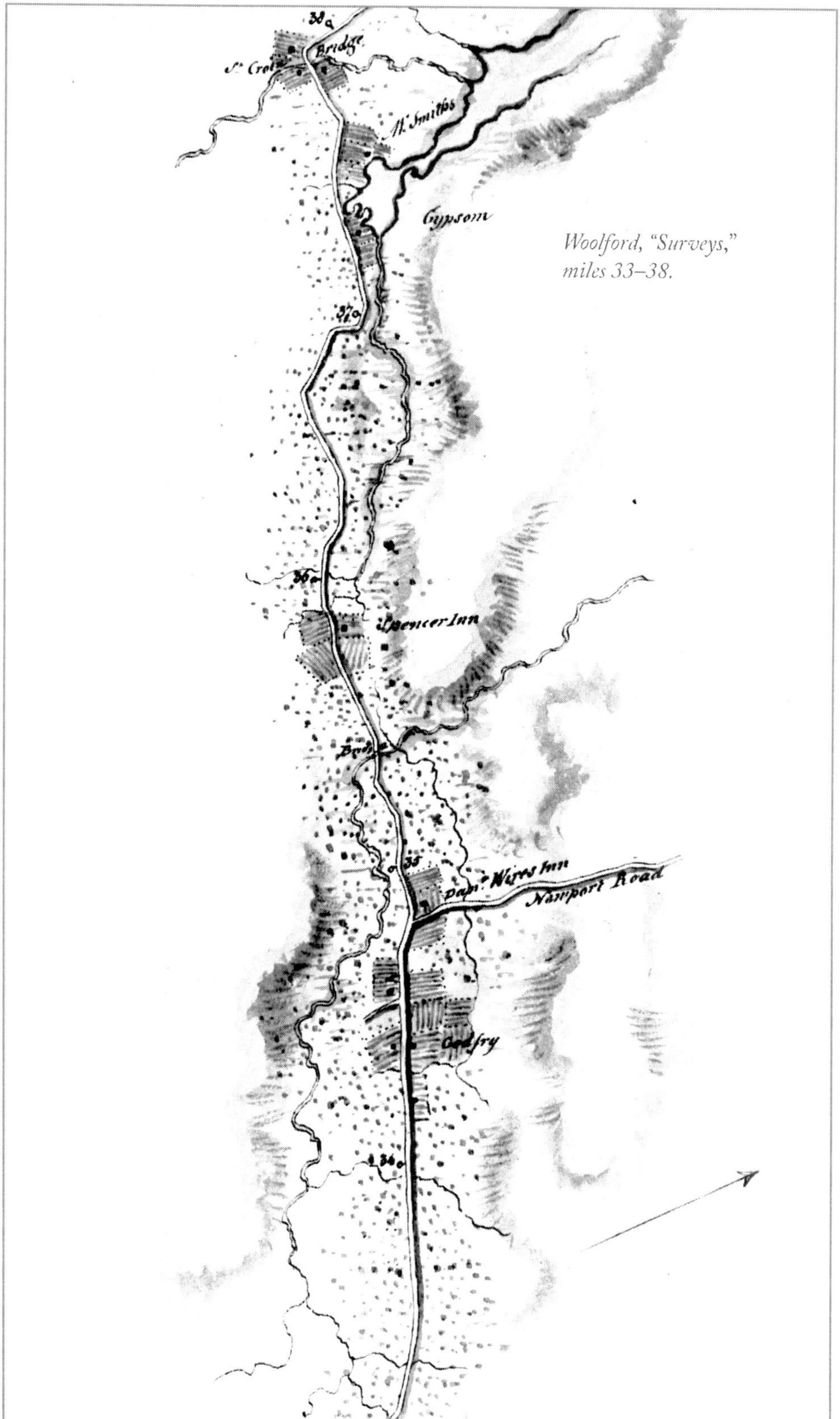

Woolford, "Surveys," miles 33–38.

Scotia, he was impressed with the setting of Morris's land, on which "a small hut and a barn" had been erected by 1817. Morris's grandfather had come to Nova Scotia in the mid-1700s and been appointed chief surveyor, a post in which he was succeeded (with the title changed to surveyor general) by his son, then his grandson (at the time the incumbent), and then his great-grandson.

Just beyond the Morris property lived another member of the Thompson family, whose small farm lay on the lower slopes of Ardice (Ardoise) Hill. This height of land, on which another unidentified property is shown, marked the transition between the rocky, fire-ravaged terrain to the south and the more fertile farmland on the Windsor side. More hardwood trees were found mixed with spruce, hemlock, and hackmatack, and the woods on each side of the road were thick and undamaged by fire. The population grew denser, with small cottages in cleared fields and with more clearing underway. Lord Dalhousie was pleased to see this development, but as a gentleman of his day who liked to observe nature in its most picturesque form, he complained, "There is nothing romantic on all this road, and when the wood is cut down, which at present is the grand object of every new settler, there will be little to interest the traveller's eye." He did, however, concede that, "the day being clear, we had an extensive view from Ardoise Hill of Windsor, Horton, Cornwallis, Cape Blowmedown [Blomidon], & the Bason of Minas towards the Cumberland district."

The road over the hill—actually Willow Hill; Ardoise is off to the right—ran straighter than Trunk 1 does today. The road was altered to provide a more level route, but its old path still exists as a back road over the hill—with a slight alteration from its original route—and along Weir's Brook. The two routes come together again at the foot of the hill, where a small cluster of houses and cleared fields had grown up by the early nineteenth century. There were several houses, but the only proprietor identified was a man named Godfrey. Just before mile thirty-five, the Newport Road turned off to the right, at what has since become known as Newport Corner, and was distinguished in 1817 by the welcome sight of Daniel Wire's (Weir's) inn. This establishment was strategically placed on the right side of the road, just beyond the intersection, to serve travellers before or after they tackled the nearby hill and also for those using the Newport Road.

A mile farther along the road, after crossing the brook, the traveller would have come to Spence's or Spencer Inn, also on the right, near the source of a tributary of the St. Croix River. Lord Dalhousie stopped there on a snowy sleigh ride from Windsor in February 1817 and was not impressed by Mr. Spencer. "We stopped a little while at a small miserable public house, kept by a man Spencer, whose only recommendation is his wife, a very pretty woman who supports a large family by her industry alone. Her story I don't yet know, but she is the daughter of a clergyman, near Falkirk in Scotland, & ran away from her parents with this man. She spoke of them with tears overflowing. They were alive last spring. Hereafter I shall endeavour to learn her story more fully." Unfortunately His Lordship does not record the rest of this sad story, if he ever discovered it. We do, however, have a sketch of the inn, produced by Woolford, showing the inn building and a cottage across the road.

The road then ran beside an arm of the St. Croix River, with glimpses of gypsum cliffs and the dyked meadows along the river to the right at mile thirty-seven. This scenery is still visible from Highway 101 today. Two homes were a short distance before the St. Croix Bridge, of which the second was identified as the home of "Mr. Smith." The traveller would have then come to the bridge itself, where on

John Elliott Woolford. "Spences [sic] *Inn on the Windsor Road." (Nova Scotia Museum, 78.45.58) The inn is the building to the right of the road.*

the left side stood the gristmill owned by Nathaniel Smith. Smith was one of the earliest British settlers in this area and had been living there for about forty years. The old mill that stands near the bridge on Highway 1 today is not the original one but a much larger building.

St. Andrews Bridge to Windsor

Minor alterations have been made over the years to the last stretch of the Windsor Road. The older route has survived in the communities on the approach to Windsor as loops off to the side, identified in places as the Old Halifax Road. Once over the steep hill beyond St. Andrews Bridge, the road passed what was then a very sparsely populated area with only two or three farms, and soon after it began to follow a brook through marshy country.

Burdain's Inn (elsewhere spelled Burdon) with its surrounding fields was located at mile forty-one, shortly before the road to Falmouth turned off to the left through Three Mile Plains. The inn, situated at the intersection of two roads, was strategically positioned to serve travellers in all directions. In those days, before there was a bridge at Windsor, travellers to Falmouth and beyond had the option of taking this road to the bridge at Upper Falmouth instead of taking a ferry from Windsor. This route, known today as the Windsor Back Road, reduced the distance of the journey to the Minas Basin and the Annapolis Valley.

After the intersection, a series of cultivated properties came into view. Most of the land along the road was by then cleared, and settlement grew denser as the traveller entered Windsor. Some of the leading citizens of the town owned property in this area. At the Falmouth turnoff stood the home of Mr. DeWolfe, a member of a prominent local family. The property on a hill a mile farther along belonged to Mr. Smith, the sheriff of Hants County. He came to the rescue of Lord Dalhousie and his companions when they were returning to Halifax by sleigh and the road was clogged with snow. Sheriff Smith brought out his men with horses to beat down the snow. His neighbour was Mr. Godfrey, whose property stood near a brook with deposits of limestone nearby.

The scenery along the final mile into Windsor was much more varied. A road ran off to the left, as it does today at what is known as Curry's Corner, to join the Falmouth Road. This is now Trunk 14, linking Windsor and Chester. On the right, the traveller could see the parish church and burial ground. The church's foundation stone had been laid in 1789 by Bishop Inglis. The old burial ground

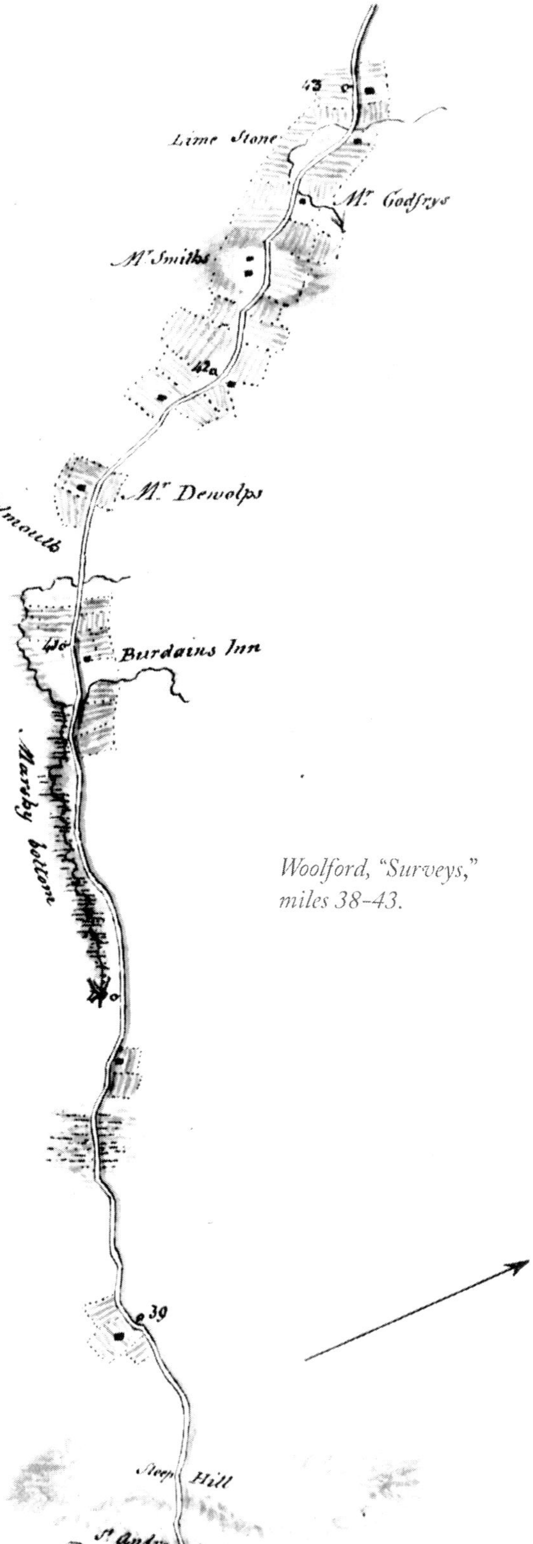

Woolford, "Surveys,"
miles 38–43.

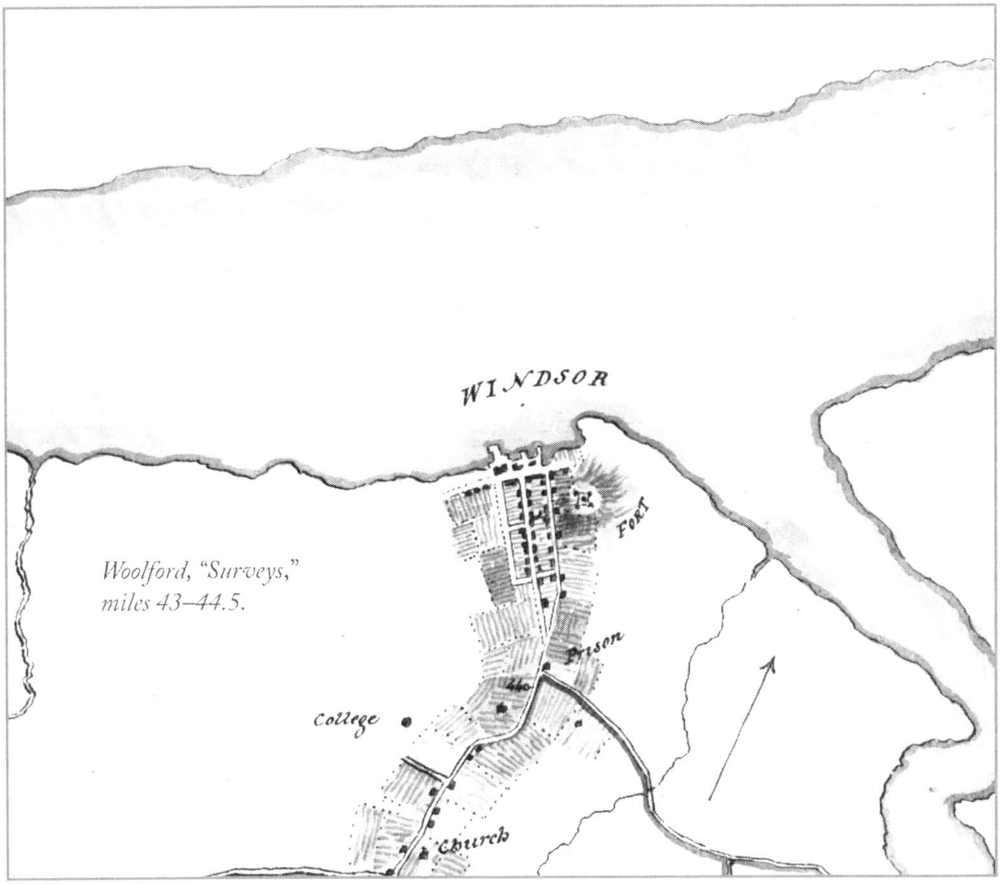

Woolford, "Surveys," miles 43–44.5.

remains at this location, but the church burned down in 1892 and was replaced by a building closer to the town's centre.

The next road to the left led to the grounds of King's College, the oldest English university in Canada and the oldest in the Commonwealth outside the United Kingdom. Marked simply "College" on the map, it had been established in 1789, and it could be seen situated high on a hill, looking very impressive from the distance. It was an exclusive Anglican establishment, supposedly based on the traditions of Oxford and Cambridge.

The time of Lord Dalhousie's administration was a low period in the college's history. His Lordship recorded in September 1817 that "the number of students at present is only fourteen. The state of the building is ruinous; extremely exposed by its situation, every wind blows thro' it. The passage doors are torn off, the rooms of the students are open and neglected." These were not the only complaints that

John Elliott Woolford. "Windsor from the E." (Nova Scotia Museum, 78.45.56) The road ran along the hollow between the parish church and King's College, both visible in this sketch.

the lieutenant-governor had to make. He disapproved of its exclusive Anglican rules in a province where non-Anglicans formed the majority of the population, and he was disturbed by the friction he observed among the faculty. King's survived the troubles of this period, but the main building was destroyed by fire in 1920. The University of King's College was rebuilt in Halifax. Its affiliated school, today's King's-Edgehill, remains on the Windsor site.

Mile forty-four was marked by a road leading off to the right and by the town's prison that stood at the intersection. The Winckworth (now Wentworth) Road led to the bank of the St. Croix River, where a ferry took travellers across to Newport. It ran past the properties of some of the leading citizens of Windsor, including Chief Justice Blowers, whose house stood to the right on the far side of the creek, and James Fraser. Fraser's house, known as the Grove, lay a little further along the road and was rented for the summer of 1817 by Lord and Lady Dalhousie.

The traveller finally came to the town of Windsor, which was laid out in a simple rectangular grid fashion, whereby the road divided as it does today to form the two main residential streets and cross streets. Each house had a long narrow yard behind, and land around the outskirts of the town was under cultivation. Dominating the town from a hill on the right was Fort Edward, whose blockhouse still stands. The waterfront, with its wharves and commercial buildings, was an important feature in those days, as shipping services linked Windsor with other settlements on the Minas Basin and the Bay of Fundy.

There was no bridge across the Avon River at Windsor until the 1830s, when a covered toll bridge was constructed. The bridge at Windsor brought more traffic to the town and made it a good starting point for the journey to Annapolis Royal, Digby, and Yarmouth. This route will be explored in a later chapter. Windsor's position on the Avon River estuary encouraged passenger transport and trade with other parts of North America by way of the Bay of Fundy, and an efficient road from Halifax was necessary to link the capital with the busy shipping port. The existence of a viable road allowed the citizens of Windsor to benefit from commercial interaction with Halifax, and also to participate in its social and cultural activities.

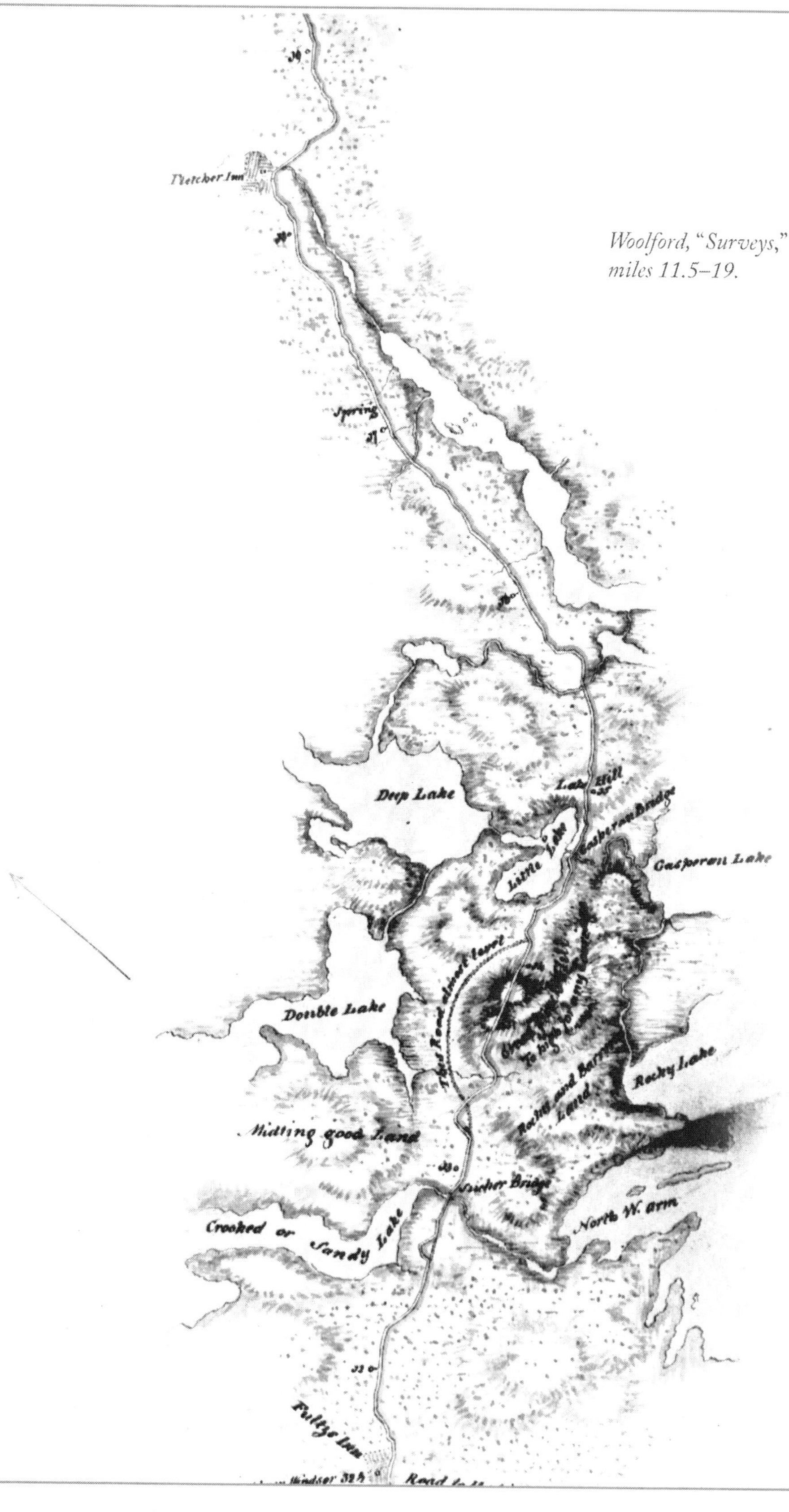

Woolford, "Surveys," miles 11.5–19.

3

The Truro Road
(Circa 1818)

The Truro Road and the Windsor Road parted company at Fultz's Inn. The road to Truro ran to the right, following the route that still exists as the Old Cobequid Road. This name comes from the Mi'kmaq name for Truro that was also used by the Acadians. By 1818, this was the route of Nova Scotia's second stagecoach service, established in 1816 between the capital and Truro. At this time, the road was not as well-settled as the Windsor Road, but it was the first link in the land route to Amherst and the border with New Brunswick, and the road to Pictou, Antigonish, and the Strait of Canso. For this reason alone it was a significant provincial roadway. Allowing for some realignment, it is still possible to trace many parts of the road Woolford mapped in 1817–18.

The early nineteenth-century road was subjected to major alterations, so that Trunk 2 follows rather a different route for much of its course. After leaving Fultz's Inn, the original road remained on the west side of Lake Thomas as far as the bridge at the south end of Fletchers Lake, where today's Trunk 2 meets the earlier route at Fall River. On the east side of Fletchers Lake and Shubenacadie Grand Lake, the old road generally ran farther away from the lakes than Trunk 2 does. The older road remained on the eastern side of the Shubenacadie River, while Trunk 2 and Highway 102 cross the river. The old road continued approximately along the present county boundary, but this stretch was abandoned and now exists only partially as a trail until it approaches Nine Mile River, where it reappears as a minor road.

With some variations, the Truro Road followed the line now taken by Route 227, continuing through Carrolls Corner as far as Gays River. From here its route can be followed, again with realignments, as a country road passing through Coldstream and Stewiacke East. Across the river, the road proceeded towards Truro on the eastern side of Shortts Lake. Trunk 2 picks up the earlier route a little north of Alton, but they diverge again from time to time, so that for much of the final stretch into Millbrook and Truro, the present route of Trunk 2 runs parallel to the old road.

From Fultz's Inn to Hall's Inn

Fultz's Inn was the last habitation that travellers passed before they set out towards Truro. In 1818 they would have travelled through very hilly, unimproved, and uninhabited woodland for seven miles between a series of lakes. Although the land around there was not yet cleared for cultivation, it had recently been assessed by surveyor Valentine Gill as "midling good [*sic*]" to the left of the road, while that on the right was "rocky and barren." The land on the left was developed later and is now part of the sprawling community of Lower Sackville.

At mile thirteen, counting from Halifax, a bridge known as Sucker Bridge, presumably because of the presence of leeches in the water, carried the road over a brook linking an arm of Rocky Lake to what was then called Crooked or Sandy Lake, today's First Lake. (Several lakes in this area have changed their names over time from their former descriptive names to less interesting ones. Second Lake was known in the early nineteenth century as Double Lake, while Third Lake was called Deep Lake.) Even before 1818 the road had already been altered from its original state to accommodate wheeled traffic. A short portion had been straightened just before the bridge, but a much greater change was proposed to avoid the dangerously steep slopes of the Great Beech Hill, which was "too high for any carriage," by providing an alternative route curving to the left and rejoining the original road at the far side of the hill. Travellers on horseback may still have preferred the shorter, steep route with a fine view from the top.

After crossing Gaspereau Bridge between Little Lake and Gaspereau Lake, (now Three Mile Lake and Powder Mill Lake), the road passed over Lake Hill, which initially had presented another challenge to wheeled vehicles. This obstacle had been overcome by taking material from the top of the hill and using the resulting earth and rock at the bottom to reduce the gradient. At the far side of the hill, the road turned north to run along the west side of a long lake (Lake Thomas today). This section of road was still unpopulated. Just after mile seventeen there was a spring, where thirsty travellers and their horses might have stopped for refreshment. If they endured their thirst for another mile, they would come to Fletcher's Inn at the south end of the lake that bears the same name. This inn, situated near a bridge over the stream linking the head of Lake Thomas with the lower end of Fletchers Lake, had been run by Robert Fletcher since about 1815 and must have been a welcome sight after seven miles of wilderness.

Across the bridge, the road remained on the eastern side of the lakes and the Shubenacadie River. Most of the old road has vanished; today's Trunk 2 runs close to the lakeshore. Today, the Old Holland Road marks part of the original

John Elliott Woolford. "Road Scene (Mr. Holland's)" *(Nova Scotia Museum, 78.45.28). The Truro Road passed by the Holland property. In the foreground, the remaining stumps indicate that clearing was ongoing. The sketch also shows the type of carriage in which individuals travelled at the time.*

Woolford, "Surveys," *miles 19–27.*

route. In this area, for about fifteen miles, much of the woodland beside this road was burnt when Lord Dalhousie passed this way, leaving "rocky and barren land with low brushwood, and fireweed in summer time." However, this section of the road showed evidence of settlement. Soon after crossing two bridges over parallel streams, travellers would come to a property on the left owned by a man called Jeffery (or Jeffrey). An inn was formerly operated at this location by Colonel Fultz. Mr. Jeffrey was identified by Lord Dalhousie as the Collector of Customs, who was developing a farm nearby at Indian Point on Grand Lake, part of Oakfield Provincial Park today. In 1815 Jeffrey had been responsible for overseeing the housing of black refugees on Melville Island in Halifax after the War of 1812.

A little farther along on the right, stood Blades's Inn on a property owned by a Mr. Holland. The Hollands were a family of printers and paper-makers who flourished in nineteenth-century Nova Scotia. The best-known, Anthony Holland, was printer, founder, and editor of the *Acadian Recorder.* He also established a paper mill at Bedford and undertook various other business enterprises. It was in the *Acadian Recorder* that Blades's advertisement ran during the summer of 1817: "The Subscriber informs the Public that he has taken that convenient Stand on the COBEQUID ROAD, belonging to Mr. Holland, and has opened a House of Entertainment. He will keep good LIQUORS, and strict attention will be paid to his STABLES. JOHN BLADES." This may be the same John Blades who died in 1867 at the age of eighty-eight and was buried in the Gays River churchyard.

Identified only in pencil on Woolford's map, the Holland property was more extensive than it appeared. It consisted not only of the house, which is on the map, but also of a barn, which presumably also served as a stable, and several cleared fields, as we see from the sketch made by Woolford in 1817. The Holland name persists on this part of the old road, which today is identified as the Old Holland Road. After a short distance, this road dwindles to a trail and then peters out. It will reappear as a minor road identified near Enfield as the Old Post Road and eventually become part of Route 277.

The Kenty house, formerly an inn, stood just before mile twenty-three, at the next intersection, where a road still runs off to the left towards Grand Lake. The main road continued through several more miles of wilderness before coming to Hall's Inn at mile twenty-six, recommended as a respectable establishment by the patronage of the lieutenant-governor. The Hall property stood near a bridge over the scenic Eight Mile Brook, at a fork in the road where the left branch turned off

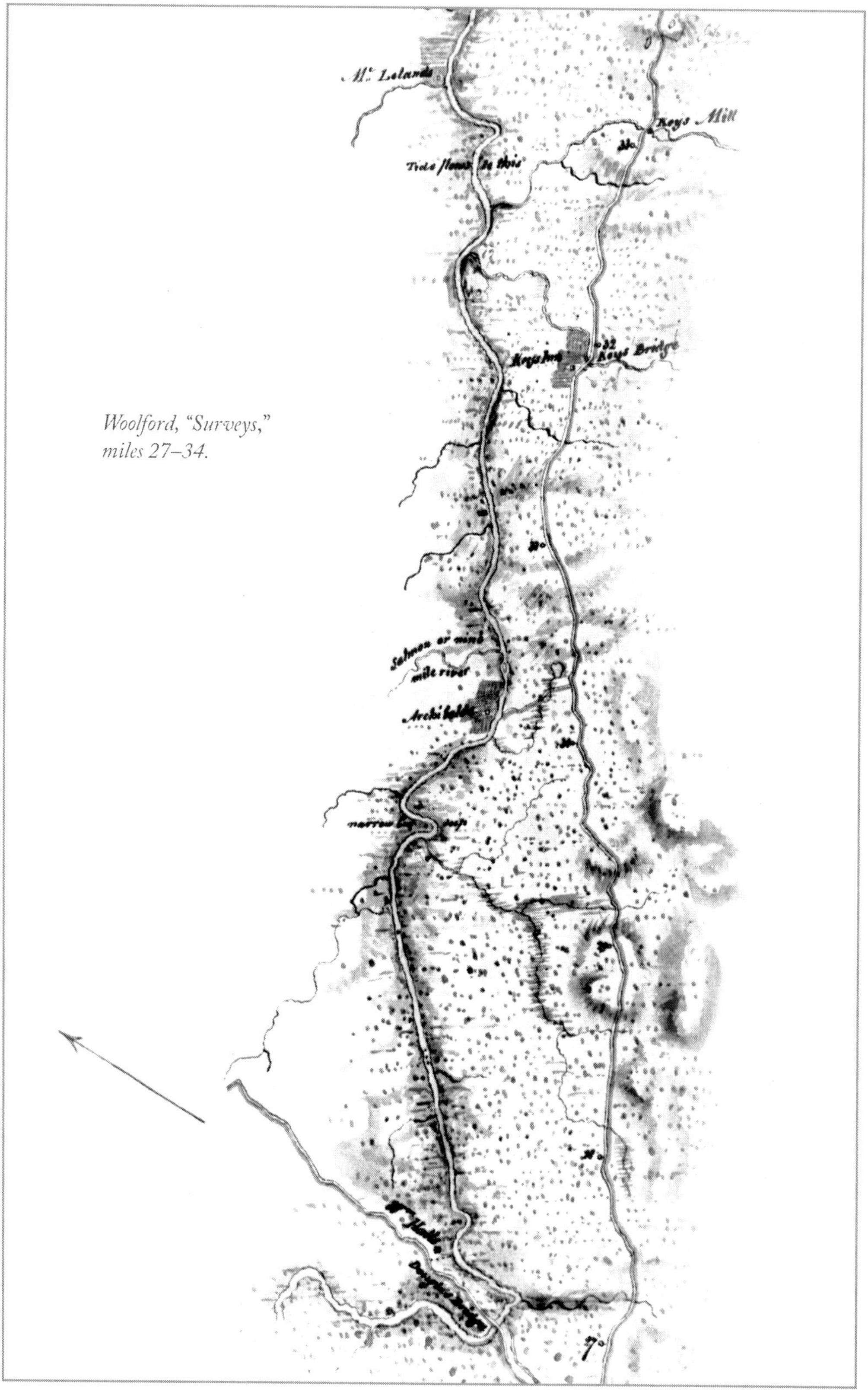

Woolford, "Surveys," miles 27–34.

towards Douglas Township in Hants County. This road is traced for a couple of miles on the next section of the map, over the Douglas Bridge and past William Hall's property on the Shubenacadie River.

Following the Shubenacadie River to Gays River

While Trunk 2 crosses the Shubenacadie River near Enfield, the old Truro Road remained on the east side of the Shubenacadie, in parts at some distance from the river. Near the Enfield approach to Highway 102, the Old Post Road is identified by a road sign. Today's Route 277 approximates part of the earlier route.

There were no dwellings on the hilly stretch of the six miles between Hall's and Keys's inns, but a road to the left just beyond mile thirty led to the riverbank opposite the Archibald property near Nine Mile River. In 1817 John Archibald was offering for sale in the *Acadian Recorder* "land and a dwelling house, sixty acres under cultivation, on river and main Cobequid Road." This was one of several properties that had been developed along the banks of the Shubenacadie. Access to many of them was probably easier by water than by land.

Keys's Inn, at mile thirty-two, was visited from time to time by Lord Dalhousie, who spoke more highly of the food and the service than of the accommodation: "miserable quarters...[but] a very good supper." The inn stood among cultivated fields to the left of the road, beside Keys's Bridge. A mile farther along the road was a mill, also run by the Keys family—the first sign of industry along this road. A millpond with a timber-built dam lay to the right of the simple bridge. The stream that powered Keys's Mill flowed westwards to join the Shubenacadie River, very close to the upper limit of the tide on that river. Some land beside the mill was being cleared for cultivation, and an extensive area of surrounding woodland had been devastated by fire. The Keys family is well represented in the Gays River graveyard; John, who died in 1867 at age seventy-two, is the oldest identified member. Just beyond the mill, the road passed between two small, round ponds named in pencil on the map as Spectacle Lakes. They are still visible as Route 277 passes through Dutch Settlement.

The next section of road served an area where the population was the most dense thus far. Many of the inhabitants, however, had established themselves on farms beside the rivers, and only a few houses were visible along the road itself. At Black Rock (mile thirty-four), the beginning of this section where the Shubenacadie River turns away to the left, the road continued in its general northeasterly

Woolford, "Surveys," miles 34–41.

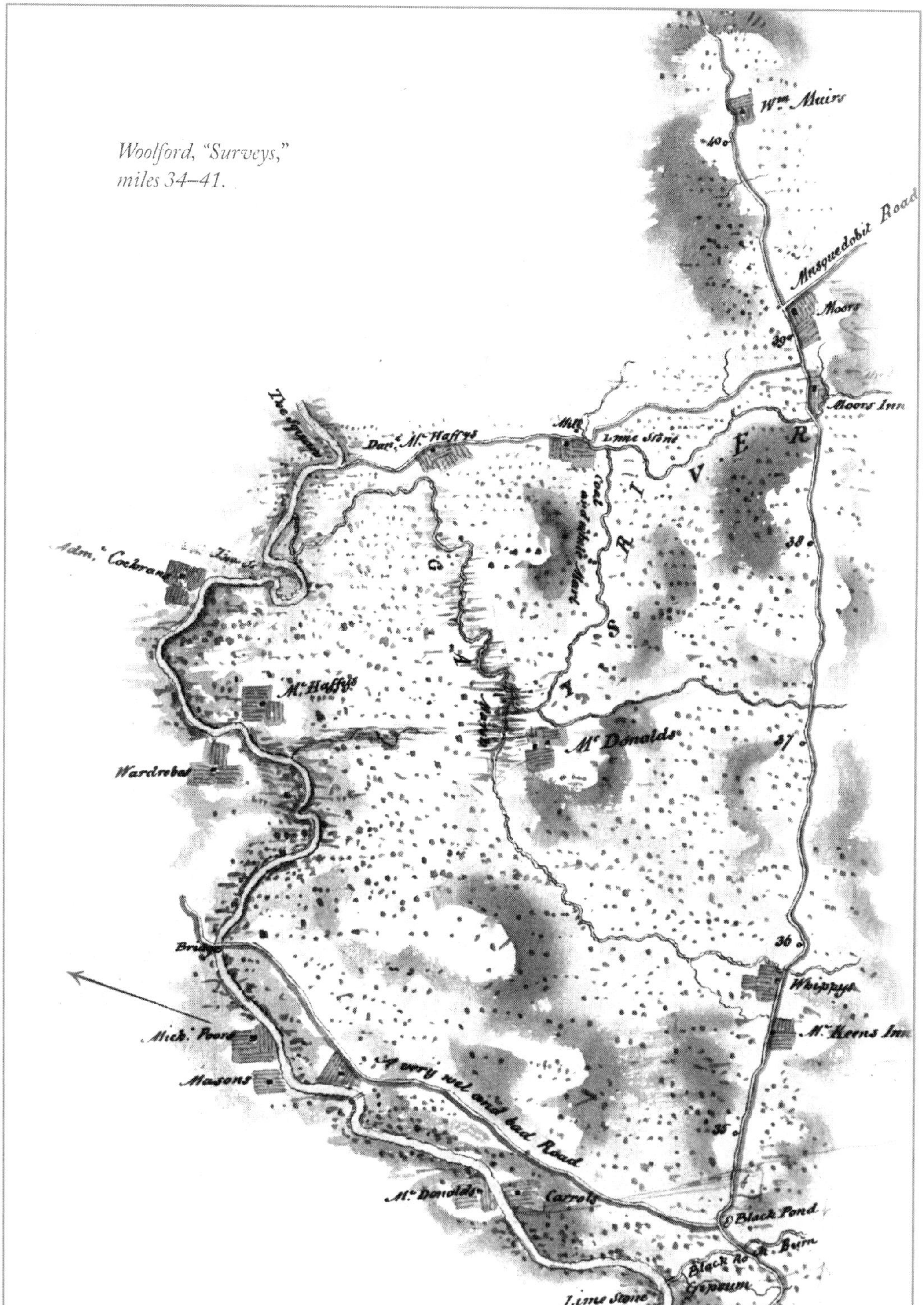

direction. Limestone and gypsum deposits were already known to exist on each side of the river here, promising future prosperity for the area. Today the National Gypsum quarry at East Milford is the largest in the world. Early-1800s travellers would have been surprised to know that the remains of a mastodon lay buried here and would be discovered in the late twentieth century.

A number of settlers had farms along the riverbank below Black Rock Burn. Some of these were accessible by what was described on the map as "a very wet and bad Road" that turned off to the left at Black Pond and crossed the Shubenacadie River by a bridge after about three miles. This road has mostly disappeared; the bridge that now crosses the river at Milford Station leaves Route 277 about two miles farther on. Several of the riverside farms could be approached only by water. These were identified as McDonald's, Mason's, Michael Poor's, Wardrobe's, McHaffy's, and the property of Admiral Cochrane. Gays River, with its tributary Big Pond Brook, meandered through the landscape between the road and the Shubenacadie River. In the middle of this area lived another McDonald family, whose land overlooked the marshy area known today as McPhee Pond. Many of these family names were still found on both sides of the Shubenacadie River when Ambrose Church created his county maps in the 1860s and 1870s.

For travellers on the main road, McKeen's Inn provided a break in the journey a mile beyond Black Pond. Beside the inn was the Whippy homestead, next to Big Pond Brook just before mile thirty-six. There were no more houses along the road until after it had crossed Gays River. The early road ran straight over a hill between Carrolls Corner and Gays River but was later replaced by a road curving round the lower slope to the south. Just across the bridge was an inn formerly run by the Moors and taken over in 1818 by the Andrews family (whose name is written in pencil on the map).

A road to the left (now Route 224) led past Daniel McHaffy's mill, which was powered by a brook running into Gays River, at what is today Pine Grove. McHaffy's home lay a little further along the road. Another road, leading to Musquodoboit, turned off to the right after a short distance, with the Moors' house and farm on the corner. The Musquodoboit Road also forms part of today's Route 244. Finally, William Muir's property at mile forty marks the end of this section of the map.

Near McHaffy's mill there was already evidence of other mineral resources: limestone, coal, and white marl are identified on the map. Marl, a form of calcium carbonate, was at one time valued as a fertilizer and had various other industrial

John Elliott Woolford. "View on Gays River (38½ Miles from Halifax on the Road to Truro)" (Nova Scotia Museum, 78.45.31). This fine truss bridge carried Truro Road travellers across Gays River.

uses. Later, gold was identified in the Gays River area. More recently, zinc and lead mining have been developed.

Although little of the settlement depicted on this map actually bordered the road, the combination of arable land and mineral resources represented in the area ensured its continued development. In 1817 there was still no obvious centre of population in this young community; however, the village of Gays River later grew to be the focal point for a wide surrounding area. Today, most of the family names that we have seen in this section can be found represented in the graveyard of the Gays River church, including some of the pioneer settlers in the area.

Mile Forty-one to the Township Line

The road beyond Gays River led in a northeasterly direction, where a minor road today leads to Stewiacke East. It passed through a group of properties situated relatively close together, in the area we now know as Coldstream. At a time when most of the settlers were fairly obscure, operating small farms, this spot was home to people with rather more varied backgrounds.

The first cottage on the left was built by a colourful character known as "Sir" John Oldmixon—the title was assumed, presumably, to impress North American society. He was an actor, descended from an English literary family in Somerset. He had amassed considerable debts from extravagant living in England, where he enjoyed the society life of the city of Bath. In order to escape his debts, in 1794 "the Beau of Bath," as he was known, immigrated with his wife Mary to the United States and settled in Philadelphia. There he took up market gardening while she continued to work on the stage. As a friend of Governor Wentworth, Sir John later acquired property in Nova Scotia and resettled there, leaving his wife behind. He received a grant of land on the Truro Road in 1812 and built this cottage where he lived, it is said, in solitude, although he had two illegitimate sons for whom he applied for a grant of land at Shubenacadie. His property included the oldest slate quarry in Nova Scotia, the Sibley Mountain or Wittenberg quarry.

Sir John could be seen travelling the road to or from Halifax on snowshoes in winter or on horseback in summer to take part in social functions at Government House or to attend to other business. In 1812 he was appointed a Justice of the Peace for the County of Halifax, which until 1835 included what are now Colchester and Pictou counties. In the summer of 1817 he again seems to have fallen on hard times. His house and lands were advertised in the *Acadian Recorder* for sale at public auction in a Chancery Court settlement of a dispute with John Lawson, a Halifax businessman. The sale of the property was held on August 5, 1817, and Sir John returned to the United States, where he died the following year. His interests had included fruit-growing, and he had introduced to the United States the peach variety that bears his name.

The next property on the left belonged to William Field, a brother of Robert Field, the well-known Nova Scotia portrait painter. William presumably came to the province with, or at about the same time as, his brother, around 1808. Robert had emigrated from England to the United States in 1794 and soon settled in Philadelphia. It was probably there that the Field brothers first became acquainted with the man who would be their neighbour in Nova Scotia, Sir John Oldmixon. Both brothers received grants of land on the Truro Road in 1812—the year Oldmixon received his grant—but only William settled there permanently with his wife and two daughters.

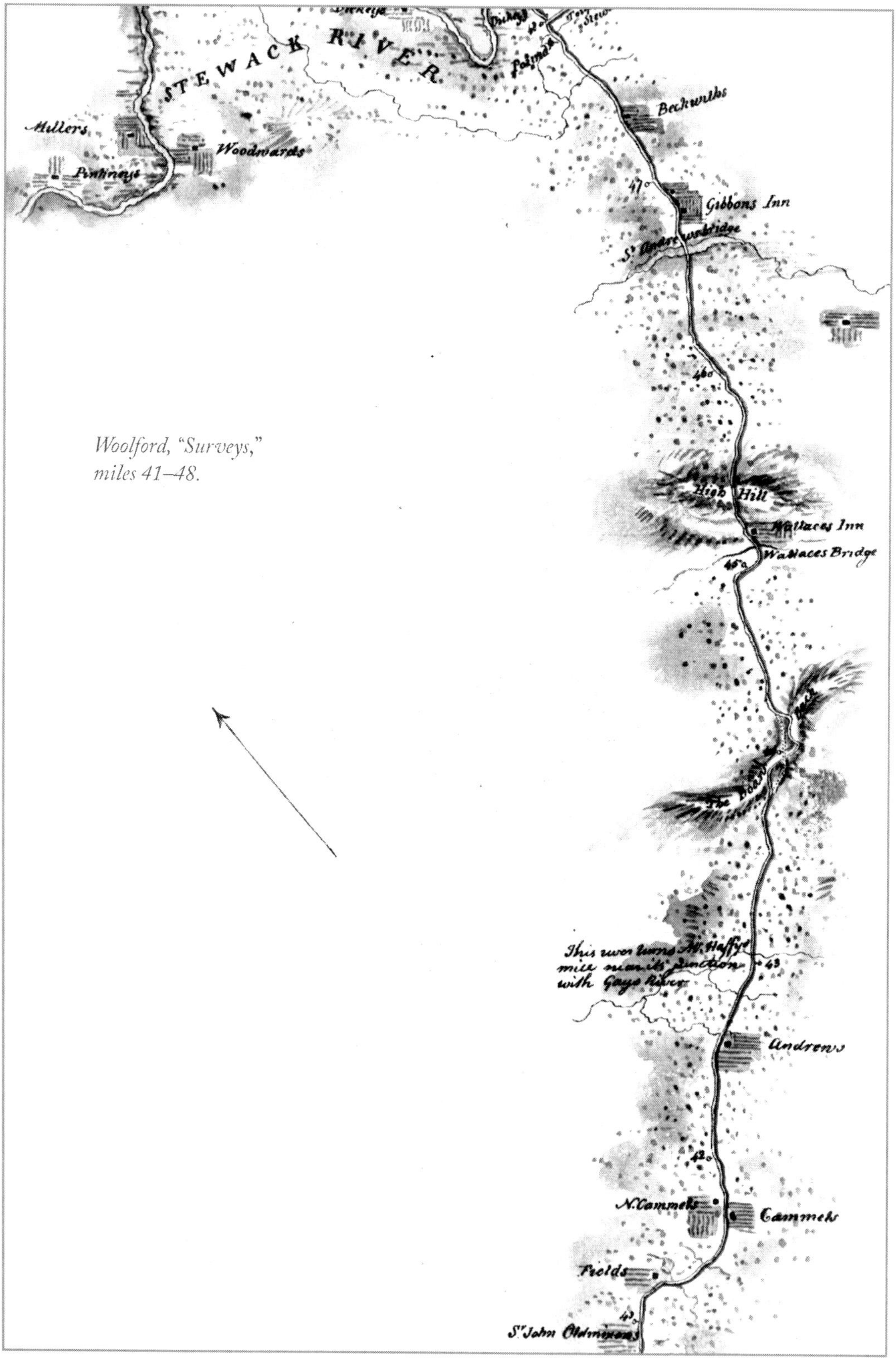

Woolford, "Surveys," miles 41–48.

Robert Field worked as a portrait painter in Halifax from 1808 to 1816 and seems not to have developed his roadside property. William Field's house fronted the road and was pleasantly situated beside a stream. It appears in a sketch by Woolford as a successful settler's property with a two-storey house, a barn with perhaps a henhouse beside it, an outhouse, neatly fenced plots, land on either side with clearing underway, and logs lying in front waiting to be sawn.

Next along the road and close neighbours of the Fields were two properties that belonged to the Cammel (perhaps Campbell?) family. About a mile farther along the road was the Andrews's farm. Just beyond the farm, the road crossed McLean Brook, identified on the map as the "river that turned McHaffy's mill near its junction with Gays River." A mile farther on, the carriage road zigzagged over a steep hill known as the Boar's Back. A more direct trail over the hill represents the original route, which was probably still used by travellers on horseback. Today's road appears to have been realigned there to the east to avoid the hill. After another mile, at Wallace's Bridge, Wallace's Inn on the right offered what must have been a welcome break in the journey. On the far side of the inn, the road ran directly over a high hill with no apparent way of easing the gradient.

At mile forty-six, a house and farm lay to the right of the road and some distance back from it. The only means of access seems to have been by following the stream that ran under St. Andrews Bridge. Just across St. Andrews Bridge, on the right side of the road, Gibbons's Inn offered another possibility of accommodation in this area. When Lord Dalhousie visited this inn on his travels, he described it as "a poor house, but very obliging people."

As the road approached the Stewiacke River, more settlement was evident both along the road and beside the river. After passing Beckwith's on the right, the traveller would have seen an enclosed area on the left side of the road, possibly holding a cow or two. This was the pound, in which animals straying from local farms were held until their owners retrieved them. Immediately after this enclosure, at mile forty-eight, the road to Upper Stewiacke turned off to the right, passing Tupper's farm after about three-quarters of a mile. A number of other farms lay along the Stewiacke River: two were owned by members of the Dickey family, others by the Woodwards, Millers, and Pinkneys, and (on the next section of the map) the Polocks and Whitemans. There is no road leading to any of these farms, and access must have been chiefly by water.

The road crossed the Stewiacke River bridge, as it does today at Stewiacke East, and passed by the two Goorly (or Gearly) homes, running northwards through an

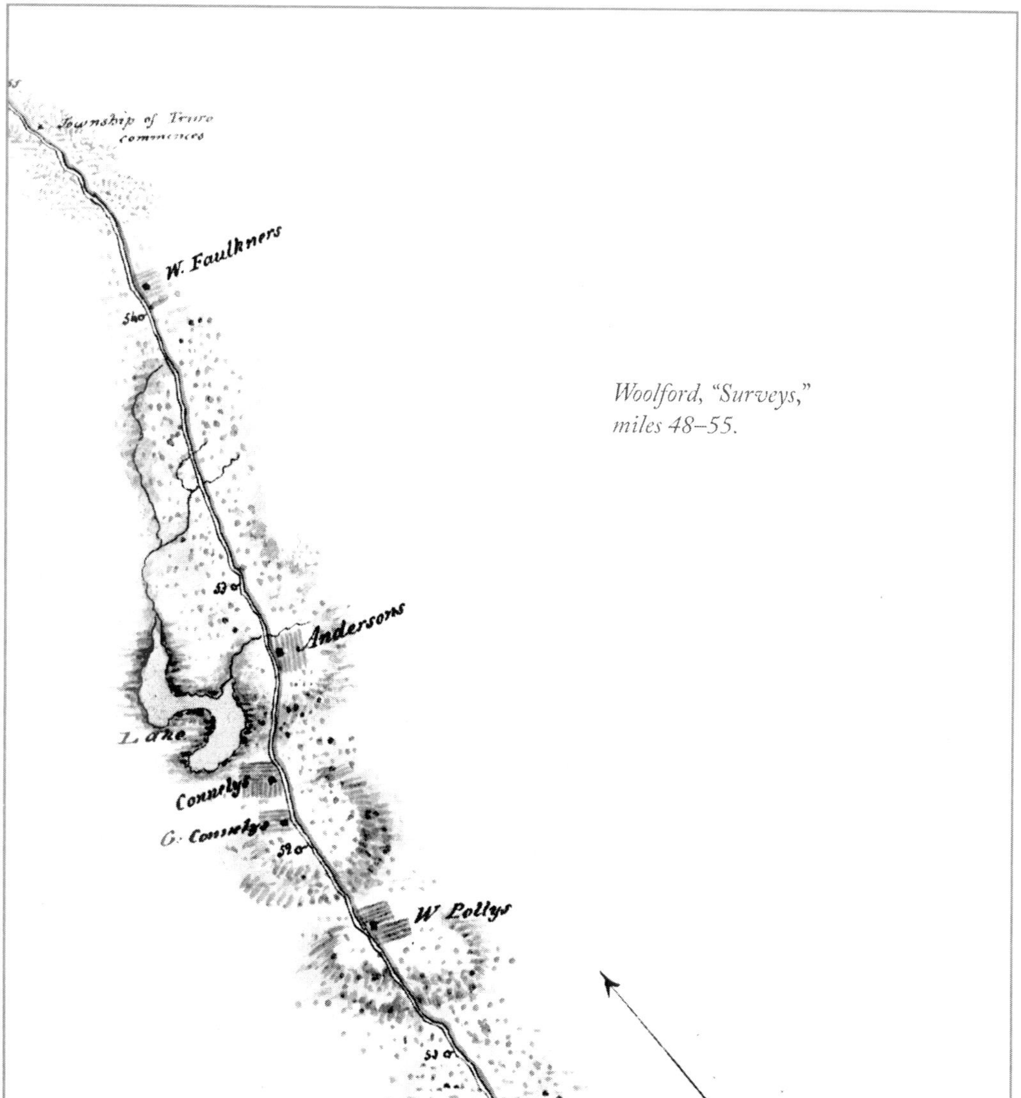

Woolford, "Surveys," miles 48–55.

undeveloped stretch before passing W. Polly's Inn, fifty-one and one-half miles from Halifax, on the right side of the road at what is today Alton. The Polly family name remains on the nearby swampy area still known as Pollys Bog. Two Connely properties stood to the left on higher ground a little farther on, just before the road passed near to the eastern side of what is today Shortts Lake. The old road ran due north from Alton and has only partially survived, while Trunk 2 now lies farther to the east, having crossed back over the river at Shubenacadie. The only other property identified on this stretch of road was W. Faulkner's home on the right at mile fifty-four, shortly before the traveller reached the boundary of Truro Township, less than a mile farther on.

The Road into Truro

The traveller passed through Truro Township into the town itself. Some of the original route runs a little to the west of the present-day Trunk 2, which has partially absorbed it. It survives as a parallel route just north of Brookfield and reappears briefly as a minor road southwest of Hilden. It stayed to the west of Mill Brook until it reached the outskirts of Truro, where it crossed the brook and ran straight along the western edge of the town.

The township was established in the 1760s by Protestant families originating in Ulster, but the rural areas developed only slowly because of disputes among the heirs of the original trustees of the grant. By 1817–18 there was still very little settlement outside the town itself. From the township line, the road ran almost straight through five miles of uninhabited woodland that Lord Dalhousie described as "a wild but beautiful wilderness," to Clark's Inn at mile fifty-nine. From there onwards, a few farms were scattered along the last few miles into the town. Some of the landholders were identified: Clark's immediate neighbour was a man named Drysdale; the Dunlops lived a mile farther along the road; and just after Truro came into view, the Wilson's and Cummins's farms stood at mile sixty-three on the final approach to the town.

After travelling sixty-two miles from Halifax, people would have been happy to reach a bend in the road from which they had their first view of their destination. Truro, formerly known by its Mi'kmaq name, Cobequid, was settled by the Acadians at the end of the seventeenth century and Aboriginal people knew it long before that. The coming wave of Scottish-Irish settlers from the United States in the 1760s began a new phase of the town's life.

Woolford, "Surveys," miles 55–63.

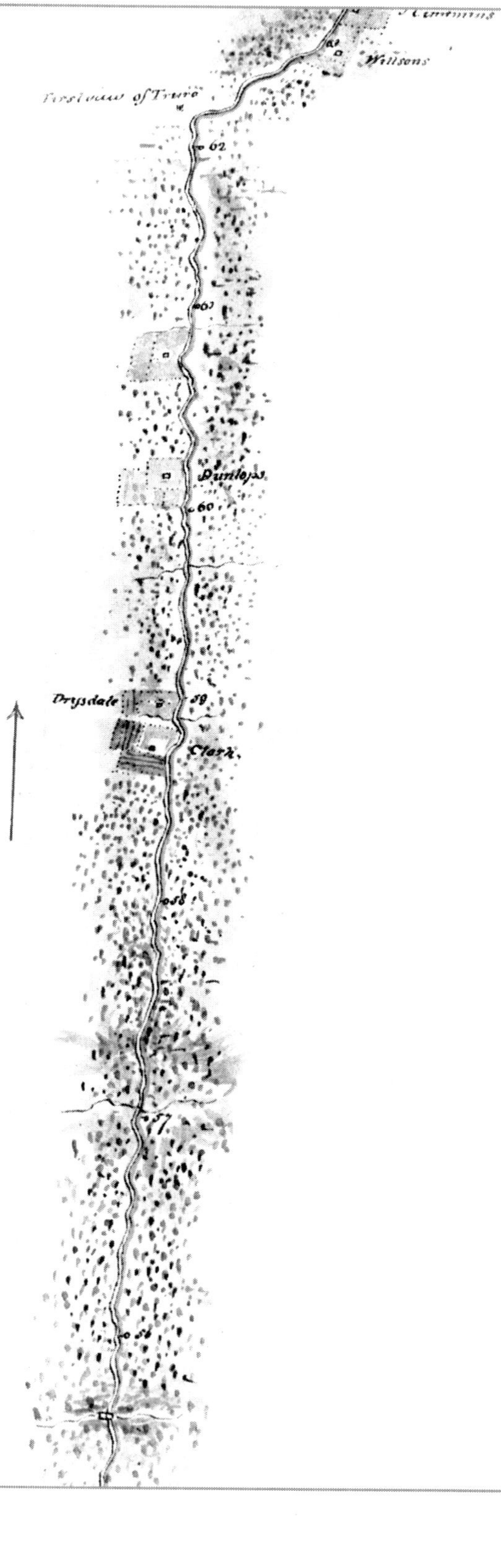

Situated at the mouth of the Salmon River and at the southern end of the portage trail to Tatamagouche, by the time this map was made Truro had already served travellers and traders for many years. It was surrounded by fertile land for agriculture and was becoming the hub of a farming community. In 1817–18, the town already had two sawmills. It was the administrative centre for the area, and its buildings included a courthouse. Also, Truro was developing into an important transportation centre, positioned at the lowest bridging point on the Salmon River with roads leading from it in all directions.

Among the first things travellers coming into town from Halifax would see was one of the town's sawmills on the outskirts of the settlement, on Mill Brook to the right of the road. The hill overlooking the small town permitted a good view of its houses, farms, and the river and hills beyond. The road from Halifax entered the town from the southwest and passed the newly built courthouse to the left, while the town itself lay to the right. Truro was laid out in the traditional eighteenth-century grid pattern. By 1817–1818, it was only partially developed, and the majority of its buildings were along the main street, which turned off to the right and ran parallel to the Salmon River. Between the main street and the river were meadows, and with the exception of that street, most of the six cleared blocks upon which the town was laid out appear to have been under cultivation, with only a few houses. There were two inns in town, Scott's and Dickson's.

By this time, Truro was already the hub of an important road network. At mile sixty-four, where the Halifax Road entered Truro, a road led to the right towards Middle Stewiacke. A proposed "road to the Lower Village," turning left at the same point, would eventually form the fourth arm of the crossroad. At the far end of the town the main road curved towards the river, where a bridge just beyond a second sawmill carried traffic to the main road leading to Pictou, a distance of about forty miles. An older, alternative route to Pictou dated from before the bridge was built. It ran on the near side of the Salmon River, crossing it a few miles upstream by means of a ford. The viability of the ford varied according to the season and the tides, but the bridge provided a reliable route for traffic to Pictou, Onslow, and beyond.

On the far bank of the Salmon River, a road led through Onslow to the ancient route along the North River to Tatamagouche. Onslow also marked the beginning of the road to the New Brunswick border near Amherst. Among other routes from Truro was the road along Cobequid Bay to the mouth of the Shubenacadie River.

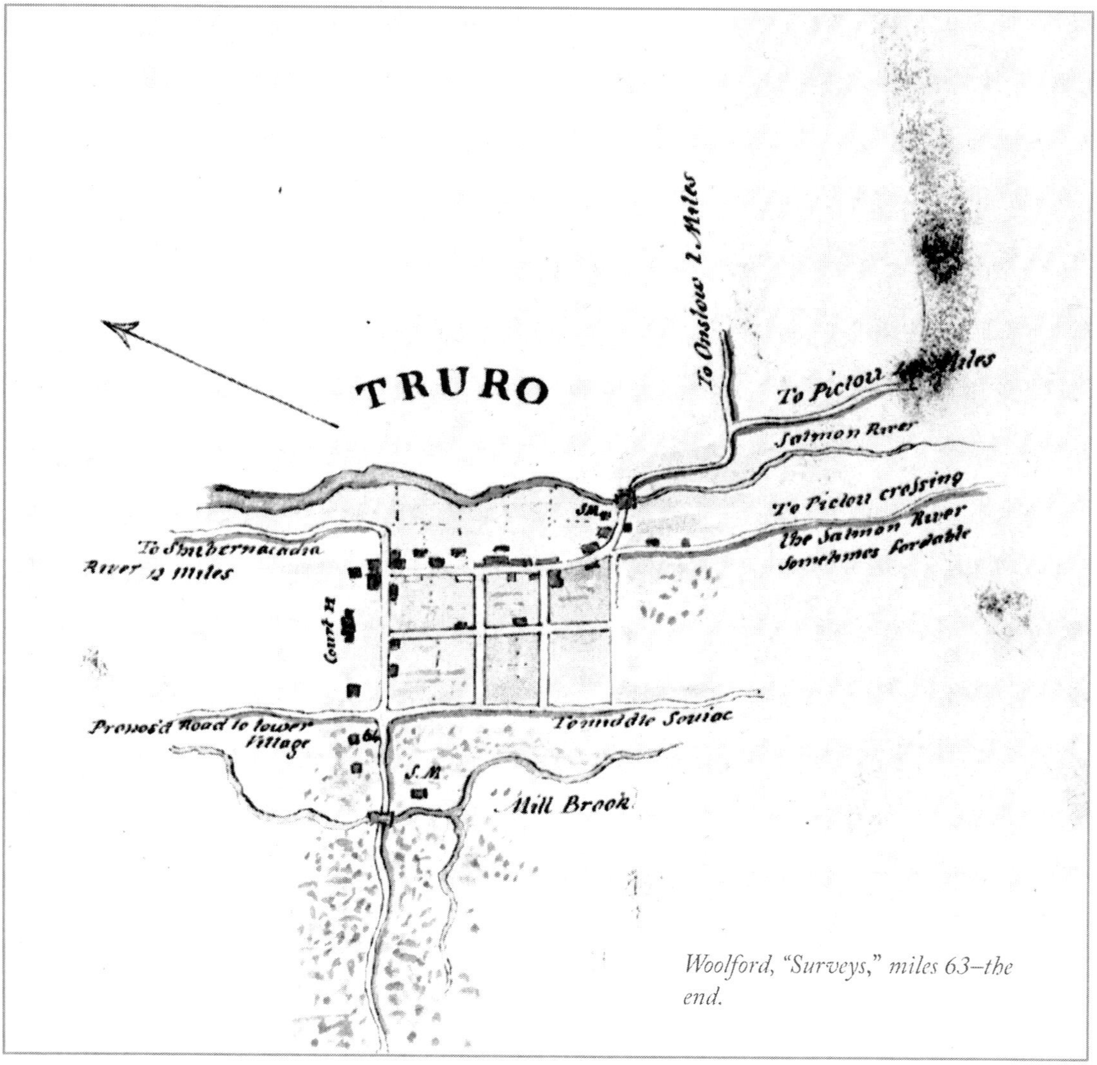

Woolford, "Surveys," miles 63–the end.

One of Truro's attractions at this time was its annual cattle show, which brought interested participants and spectators into town from as far away as Halifax. On these occasions, and on the days when court was in session, the roads into this small town must have been quite lively. Truro was the destination of one of the early stagecoach services in Nova Scotia, and it was soon supplemented by an extension to Pictou. We shall look at the development of the Pictou Road in chapter five.

But first, let us return to Halifax and its links with communities in western Nova Scotia. Here, a network of routes was already in place by the early years of the nineteenth century, though it would be some time before they could be considered practical for most travellers.

4

Western Nova Scotia

An 1817 map shows a road leading from Halifax that links the heads of many of the bays along the South Shore as far as Shelburne, cuts across country to Argyle, rounds the estuary of the Tusket River to Yarmouth, runs parallel to St. Mary's Bay to Digby and Annapolis, then follows the ancient route along the Annapolis Valley to Windsor. As well as this coastal circuit, there was a road across the province from Annapolis to Liverpool with a branch to Lunenburg, one from Windsor to Chester, many small roads in the Minas Basin area, and the line of a newly surveyed direct road designed to link Annapolis and Halifax. The appearance of these roads on this small-scale map was no guarantee of their condition, nor were their routes necessarily exactly as depicted, but we can see the beginning of a network of overland communications that was developing during the first quarter of the century. There was, in theory at least, a complete route around the western end of the peninsula serving most settled coastal areas, as well as some roads serving inland areas of the province.

A direct road from Annapolis to Halifax, planned in the late eighteenth century, was never completed at the eastern end; however, the other main roads in western Nova Scotia fared better. Trunks 1 and 3 follow close to the original routes in many areas, and where "improvements" have altered their paths, sections of the older roads can often still be traced and are sometimes identified by signage.

Highways 101 and 103 have replaced and sometimes absorbed Trunks 1 and 3 as the main thoroughfares, but much of the general road system established in the early nineteenth century has survived.

The South Shore

The development of the route along the South Shore in the early nineteenth century was slow and difficult. It is represented in a number of maps, as surveyors worked on establishing and improving links between Halifax and Lunenburg, and with the Planter and Loyalist settlements further down the coast.

It might be expected that since Lunenburg was settled so soon after the founding of Halifax, a road linking the two would have been established fairly early. In fact, travel between the two places was at first almost exclusively by water, and the land route was not much more than a trail. This route appears on a map made soon after the settlement of Lunenburg, but nearly thirty years later there was still no proper road from the capital. A new survey was conducted in 1782. At that time, the surveyor grandly identified his route from Halifax as the Public Shore Post Road, but it was little more than a blazed track along which a horse could be ridden on a good day. The route initially began some distance inland, and left what is today the Bedford Highway at Rockingham to pass over Geizers Hill and follow close to the shore of Pockwock Lake. The route then followed an almost straight line to the Mushamush River above Mahone Bay, before it turned towards Lunenburg. Today's Kearney Lake Road follows the approximate line of the older route as far as the Hammonds Plains Road. A road still runs through Upper Hammonds Plains to Pockwock, but it was the route through Armdale to the Head of St. Margarets Bay that was eventually developed in the 1840s.

By the early nineteenth century, the road that branched away from the Bedford Basin at the mouth of the Salmon River (Mill Cove) and through Hammonds Plains to the Head of St. Margarets Bay was a recognized land route out of Halifax towards the South Shore. Another road ran from the Northwest Arm westwards through today's Spryfield to the Seabright area of St. Margarets Bay, where it picked up a shore road to join the Hammonds Plains Road at the head of the bay. This route exists today chiefly as a walking trail from Spryfield and Goodwood to St. Margarets Bay, where a short section of road emerges between Glen Margaret and Seabright. The road westwards from St. Margarets Bay to Chester was distinctly unpopular with travellers. William Moorsom, who travelled over this route in the late 1820s, described it as "a miserable horse-path." Not surprisingly, many people preferred to make the journey by sea.

As a result of the poor condition of this road, for much of the first half of the nineteenth century the official stagecoach route between Halifax and Chester was by way of Windsor and along the road which became Route 14. In the fall of 1845 a new road was completed through Armdale that allowed travellers to go directly from the capital to Chester and beyond. In the July 29, 1945 issue of the *Halifax Times*, a letter from Lunenburg eagerly anticipated the opening of the road: "The barrier that existed between us will be removed—the Chester Road by Margarets Bay will be opened.... Then you may breakfast in Halifax, dine in Chester and... take your tea...in Lunenburg."

Because of the deep bays and long estuaries along the South Shore, the road that was developed westwards from Halifax linked the settlements at the heads of the inlets and the crossing places of the rivers rather than following the coastline. From Shelburne, it ran overland to Argyle and Yarmouth, leaving the Cape Sable area isolated for many years. William Moorsom found the section between Liverpool and Shelburne barely passable in the late 1820s: "The natives described this to be a 'post-road' to Halifax; I was therefore simple enough to ride my own mare; but any man who has a regard for his cattle would never trust his own cow upon such paths." Moorsom found the road between Shelburne and Yarmouth slightly better but still not advisable for wheeled vehicles: "An associate judge of the Supreme Court very nearly broke his neck not many months ago, in consequence of the bad state of the Barrington roads." By 1834 the road between Shelburne and Yarmouth had been improved sufficiently for the establishment of a weekly stagecoach service.

These roads were modified during the nineteenth century as both coastal and inland communities developed and local needs evolved. By the end of the century the route of the Post Road along the South Shore differed in many details from its original path. Some sections of the older road have survived as part of Trunk 3 and others as relatively minor roads which are pleasant alternatives to today's more direct Highway 103.

The Road from Mahone Bay to Liverpool, mid-1820s

In the late eighteenth century, maps showed land routes between Halifax and Lunenburg and from Liverpool through Shelburne to Argyle. These were not necessarily easily travelled but had at least been marked out. Although not even a trail was shown between Lunenburg and Liverpool, this gap in the land route was soon filled. Towards the end of the eighteenth century, Joseph Pernette built a road to Lunenburg from a point across the LaHave River from his West LaHave homestead. It, too, was little more than a bridle path, but it became the first section of the road that linked Lunenburg to Liverpool. Known today as the Grimm Road, its route remains largely unchanged. From the west side of the LaHave River, the road was continued through Mount Pleasant, Petite Rivière, and Mill Village to Liverpool, forming a section of the South Shore Post Road.

In the 1820s changes were made as part of Governor Kempt's plan for a major upgrade of Nova Scotia's roads. George Wightman was one of the surveyors associated with improvements made in the second quarter of the century as a result of Kempt's initiative. His map, made in about 1825, shows both the original Post Road to Liverpool with suggested improvements and an alternative route from Mahone Bay to the LaHave River. It also shows the inns along the way, beginning with Zwickers Inn at Mahone Bay. This former inn can still be seen on Main Street, surrounded by more recent buildings.

John Elliott Woolford. "Zwickers Inn, Mahone Bay from the N.W." 1818. This inn was a holiday retreat for Lady Dalhousie and her entourage in the summer of 1818.

This section of road between Mount Pleasant and Petite Rivière has changed little from Wightman's day. J. Dawson photo.

In the early nineteenth century there was no bridge across the LaHave River. Travellers from Halifax to Liverpool and beyond would have proceeded from Mahone Bay via Lunenburg. From there, they followed Pernette's road to the river, reaching the east bank of the LaHave River at what is today Middle LaHave. From there a ferry was run for many years by John Pernette, Joseph's third son, conveying passengers, including the postman and his horse, across the river. The ferry was based at Pernette's Cove in West LaHave. Pernette received a subsidy from the government to run this ferry, which was an integral part of the route along the South Shore.

On the west side of the LaHave River, the Post Road, which is identified today by a highway sign, climbed up the steep riverbank and ran back across country to Petite Rivière through what is now Mount Pleasant. A road (disused today) from a little further south, where a second ferry came ashore at Millers Point, joined the Post Road at Millers Pond.

The bridge at Petite Rivière crossed the river a little upstream from the present crossing, and the road ran briefly down the western side of the river, turning sharply up the hill towards Broad Cove. This old highway can still be followed today; the unpaved sections between the LaHave River and Route 331 and from Mount Pleasant to Petite Rivière, appear to have changed very little over the

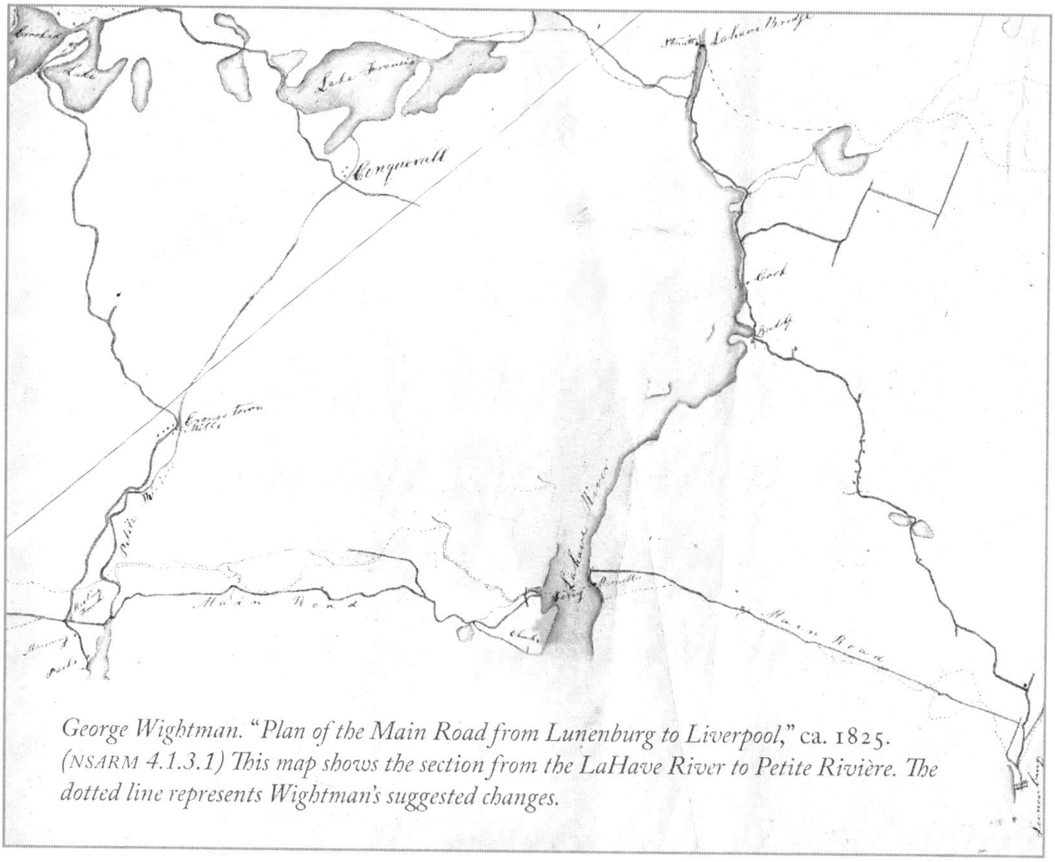

George Wightman. "Plan of the Main Road from Lunenburg to Liverpool," ca. 1825. (NSARM *4.1.3.1) This map shows the section from the LaHave River to Petite Rivière. The dotted line represents Wightman's suggested changes.*

past two hundred years. Wightman's suggested improvements were not carried out, presumably because once the "Main Road" was rerouted this section lost its importance and became a minor local road.

By the early nineteenth century Petite Rivière was a well-established settlement, with a meeting house at the crossroads at the top of the hill and two inns. The road running north from the meeting house (Upper Road today) followed the river to the mills at Crousetown and beyond. The road running down into the village (Drews Hill) offered a choice of two inns. One, the first inn in the community, was opened in 1815 by Mary Drew with the help of her sons after the death of her husband Lemuel. She married again in 1818 and became Mrs. Walter Manning. Widowed again two years later, she continued to operate the inn, then known as "Mrs. Manning's," which had an excellent reputation among travellers. As well, Mrs. Manning's inn became a Post House. The second inn was run by Joseph Park, a prominent community member who also served as a Justice of the Peace and whose family members were among the first British settlers in Petite Rivière.

Leaving Petite Rivière, the Main Road as marked on Wightman's map ran in an almost straight line towards Broad Cove, much as Route 331 does today, and then ran inland along the route known today as the "back road" behind Cherry Hill to the head of Conrads (now Voglers) Cove. The road continued to Mill Village, so called because of the mills established there by the first settlers. Just before the village, Mrs. Mack's inn offered a welcome stopping place near an intersection where a road from the interior joined the main highway. The Macks were a well-established family in Mill Village area. Mrs. Mack's husband would have been a descendant of Samuel Mack, who came to Mill Village from Connecticut in 1764 and purchased property, including the rights to the mills.

From the village, the road continued to Liverpool by a route that would become Trunk 3 and which has today been partly absorbed into Highway 103. Two sections of the older route have survived: one is a road southwestwards from Mill Village that stops short at Highway 101; the other is Trunk 3 from Exit 18 towards Brooklyn and through Sandy Cove to Liverpool Bridge. Wightman's proposed changes to the road as it neared Liverpool do not appear to have been carried out, and the two sharp angles in the route of the original Main Road from Mill Village to Liverpool can still be seen today.

The most important recommendation of Wightman's survey was the development of an alternate route "by which the distance from Liverpool to Halifax is lessened and the LaHave Ferry avoided." This was possible because of the construction of a bridge over the LaHave in 1825. Wightman's proposed new route would bypass Lunenburg, the LaHave ferry, and Petite Rivière. It cut across country from Mahone Bay, ignoring the existing grid of minor roads that served the early Lunenburg settlers' farm lots. It crossed the Blockhouse-Lunenburg road, passing north of Covey Lake towards the newly built bridge. This is roughly the path followed today by Route 325.

The location of "Lahave Bridge" on the map is of course the present-day town of Bridgewater. James Starratt's inn on the west side of the river was already in place and appears on the map. He built the inn and a wharf soon after the construction of the bridge, which stood close to the site of the newer of today's town bridges. The new route spurred the development of the town, incorporated in 1899. The road shown on the map between Lunenburg and the new bridge would become today's Lunenburg-Bridgewater road with minor alterations.

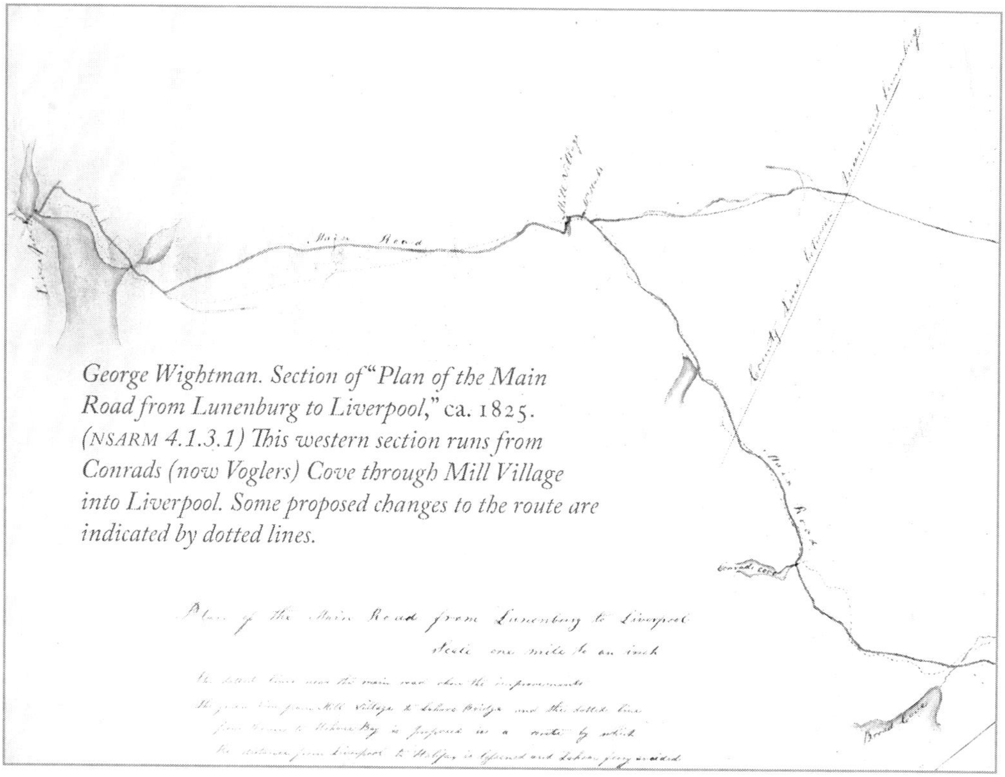

George Wightman. Section of "Plan of the Main Road from Lunenburg to Liverpool," ca. 1825. *(*NSARM *4.1.3.1) This western section runs from Conrads (now Voglers) Cove through Mill Village into Liverpool. Some proposed changes to the route are indicated by dotted lines.*

Once across the bridge, the proposed road was to run a little way down the west side of the river, as it does today, before turning off towards its ultimate destination. A road already ran westwards from what would become Bridgewater's main street towards Hebbs Mills (now Hebbville), but Wightman proposed a route a little farther north for the main highway, shortening the journey slightly, by what is now Dufferin Street. From Hebbs Mills, the survey shows the projected road passing west of "Lake Francis," which is clearly recognizable as Fancy Lake. From Conquerall, Wightman's route would pass west of Crooked Lake to Camperdown to pick up the existing road, which joined the old Main Road near Mrs. Mack's inn. This section was only partially developed according to Wightman's plan; instead, the "Post Road to Liverpool" that appeared on Ambrose Church's map of Lunenburg County in the 1880s eventually passed through Italy Cross. This road became Trunk 3, and parts of it were later absorbed by Highway 103. Sections of Wightman's proposed route can still be found in the area around Laconia and Lapland.

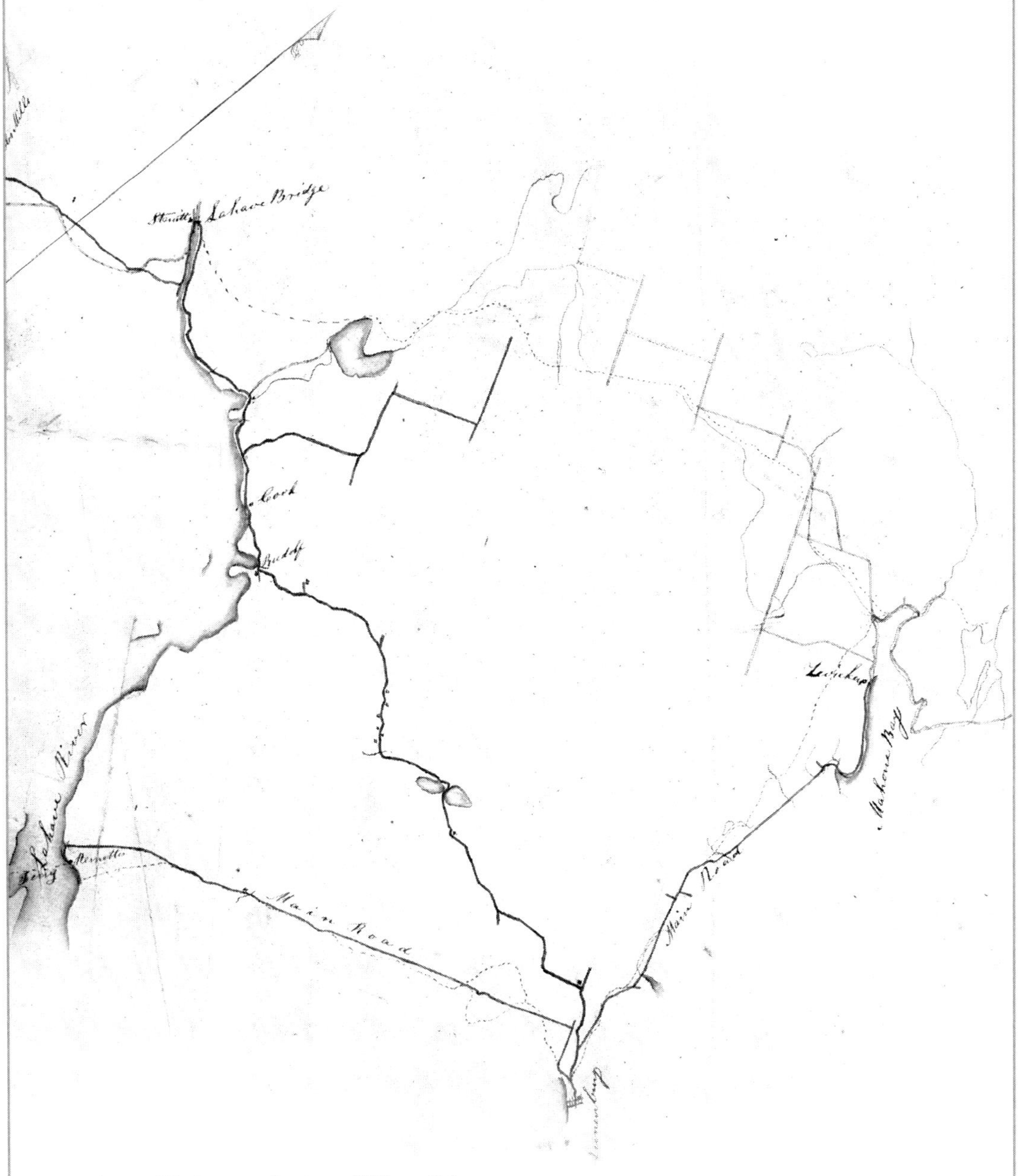

George Wightman. Section of "Plan of the Main Road from Lunenburg to Liverpool," ca. 1825. *(NSARM 4.1.3.1) The map shows how a road from Mahone Bay to the new bridge over the LaHave River would shorten the journey from Halifax.*

Although some details of Wightman's suggested changes were not implemented, his plan for a new route by way of the bridge over the LaHave River represented a major improvement to travel along the South Shore. The direct road from Mahone Bay to the bridge and the new route to Mill Village did, indeed, cut considerable distance from the journey. At the same time, they triggered the development of Bridgewater as an important transportation centre for Lunenburg County.

Halifax to Annapolis and Yarmouth

The area between Windsor and Annapolis is the oldest continuously settled part of Nova Scotia. As we have seen, the road from Halifax to Windsor followed an ancient trail across the province. The road that continued west of Windsor also followed a very early route linking the Acadian villages around the Minas Basin with the earlier settlements on the Annapolis River and Basin. The road ran south of the Cornwallis River to that river's head near what is now Kentville. It then turned northwards before following the northern side of the Annapolis River through Aylesford, Wilmot, and Lawrencetown to Granville, where a ferry crossed to Annapolis. In the early days, another ferry operated farther up the river at a spot known as Hicks's Ferry, giving access to a parallel route to Annapolis on the south side of the river. The ferry was replaced by a covered bridge in 1805, and the settlement became known as Bridgetown.

A network of small roads north of the Cornwallis River joins communities that date back to the Acadian and Planter settlers. The main road followed the rivers, as Trunk 1 does today, but the "back road" (Route 221) from Kingsport along the lower slope of the North Mountain and the "Ridge Road" along the South Mountain also represent early routes leading westwards. There are still many parallel routes among the slopes and ridges of the North and South mountains where the higher elevation afforded some protection from flooding. Highway 101 has superseded them all as the main route westwards from Windsor.

From the head of the Annapolis Basin, travellers and mail were initially transported by water to Digby and then to Saint John, making the Annapolis Valley an important route from the capital to New Brunswick and beyond. Post Houses were established at Halifax, Windsor, Horton, Annapolis, and Digby to collect and distribute mail. By the first decade of the nineteenth century, the road had been extended to Digby and Yarmouth, and a ferry was established to cross the

M. G. Hall. "View of Digby-Nova Scotia." This sketch depicts the main road into Digby in the early 1830s. Hooper, Saint John, NB, 1835.

Bear River. Although a bridge had been recommended here, it would not be built until 1864.

A weekly mail service started in 1812 between Digby and Yarmouth supplied by a courier on horseback. When a committee of the assembly examined the state of postal communications in 1815, it recommended that a wheeled vehicle carry the mail as well as passengers as far as Annapolis and Digby. The following year, Isaiah Smith petitioned the House for a subsidy to run such a service. Initially the road beyond Windsor was in such bad condition that coaches ran only as far as that town. In 1828 the Western Stage Coach service was inaugurated along the improved Post Road, carrying passengers and mail three times a week from Halifax to Annapolis by way of Windsor. (This service was reduced to twice a week for a while, but the third trip resumed in 1841.) One coach per week continued as far as Digby. In 1836 a weekly stagecoach service was started between Digby and Yarmouth, completing the route to the southwestern end of the province.

In the summer and fall of 1849 the famous Pony Express ran for a few months on a road along the north shore of the Annapolis Basin to Digby Gut. News dispatches from Halifax were carried by courier on horseback through a series of stations for fresh mounts en route and transferred by ship to Saint John for telegraph transmission to the eastern United States. The extension of telegraph service to Halifax ended the importance of that road.

The Route from Windsor to the Township of Horton, 1800-1836

Until the end of the eighteenth century, a ferry and ford at Windsor were the only means of crossing the Avon River, and they were usable only at low tide. The construction in 1795 of a bridge across the river at Upper Falmouth allowed the development of an alternative route to Horton Township and beyond. A road turning off from the Halifax-Windsor Road at Three Mile Plains gave access to the bridge. The road then ran cross-country from Falmouth to Horton Township, the area that now includes Hortonville, Grand Pré, and Wolfville. A map made in 1800 by John Shey represents this new road on which several other surveyors had also worked.

This road, more direct than the Windsor route, ran straight ahead from the bridge across what was known as Horton Mountain. When Lord Dalhousie visited the Township of Horton in August 1817, he described it as "a beautiful though very hilly road, seventeen miles in which are three distinct ranges of hills, the first rising from Avon Bridge & descending to Halfway river, next rising very steep & descending to Gaspereaux River, then rising and descending very steep on Horton."

The bridge over the Gaspereau River was known as Bishops' Bridge, and C. Bishop, who owned a farm on the north side of the bridge, was one of the new road's surveyors. The Bishops were among the Planters who settled in the Minas Basin area and were well established by the beginning of the nineteenth century, both in the Gaspereau Valley and along the Cornwallis River at what is today Greenwich, where their descendants still farm.

At the time of Lord Dalhousie's first visit in 1817, Horton boasted two inns, Graham's and Fowler's, as well as what he described as "a handsome Anabaptist Church" and a smaller Anglican church. Fowler's Inn had served as the Horton Post House from at least 1787, when John Fowler was listed as the proprietor.

His Lordship was not alone in commenting on the steepness of the "Horton Mountains." Other early travellers also described the hilly road. Joseph Howe,

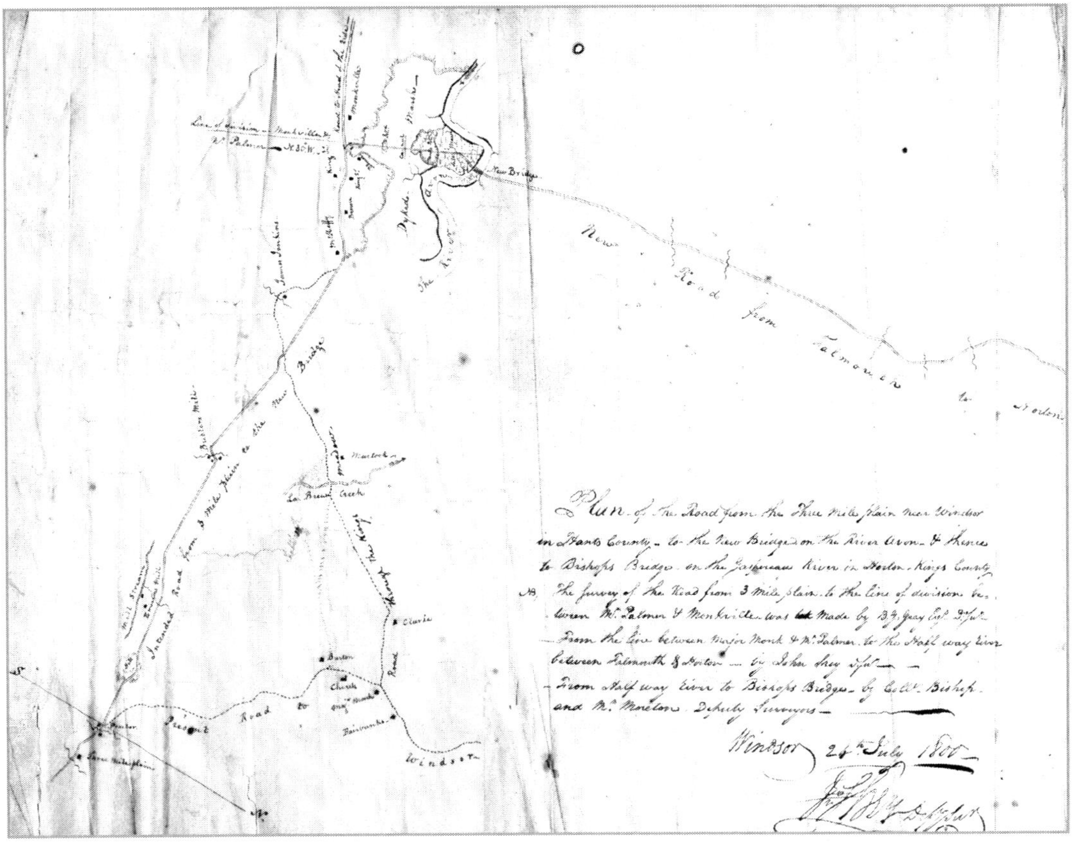

John Shey [and others]. Section of "Plan of the Road from Three Mile Plain...to the New Bridge on the River Avon & thence to Bishops Bridge on the Gaspereau River," 1800. (NSARM 4.1.2.1) The map shows the western portion, from Three Mile Plains to the bridge at Upper Falmouth, which was for several years the main Post Road to the Annapolis Valley.

travelling in the area just over a decade later, mentions getting out of his conveyance and walking up the hill to spare the horses.

On a second visit later in 1817, the lieutenant-governor left the other members of his party, including Lady Dalhousie, to pursue the direct road to Horton with a carriage and gigs, while he and his horse "turned off to the right at Falmouth Church taking the lower or Mount Denson Road which keeps by the river, & joins the other close to Horton." This road passed beside dyked farmland where cattle and sheep were fattened for market. Near Lower Falmouth, it joined the road leading from the ferry at Windsor to form the Mount Denson Road. The nine miles to Halfway River (at Hantsport) passed through "good woods, and pretty scenery" before running up a steep hill on the far side of the valley. The six miles to Gaspereau Bridge were "thickly and beautifully wooded," with only a few poor-looking clearings. The last part of his ride, from the Gaspereau Bridge to Graham's Tavern, took him by the rich farmland of Grand Pré. This road through Mount Denson, he observed, was "much longer than that over the mountain, [was] much neglected, & therefore much worse than the other."

While the road from Three Mile Plains was a shortcut for travellers from Halifax, it was not very convenient for people travelling by coach between Windsor and the Annapolis Valley before 1836. In order to cross the Avon by bridge, the coach had to make the long detour over the steep slopes of the Horton Mountains, adding many miles to the journey.

The long journey would change with the opening of a toll bridge at Windsor in 1836. The bridge provided an alternative route and replaced the ford and ferry that had linked the town with Lower Falmouth for many years. It was a

William Moorsom. Cape Blow-me-down, and the Bason of Mines, from the Horton Mountain. *(NSARM.) Moorsom's view shows the road dipping down to Halfway River as it appeared in the late 1820s.*

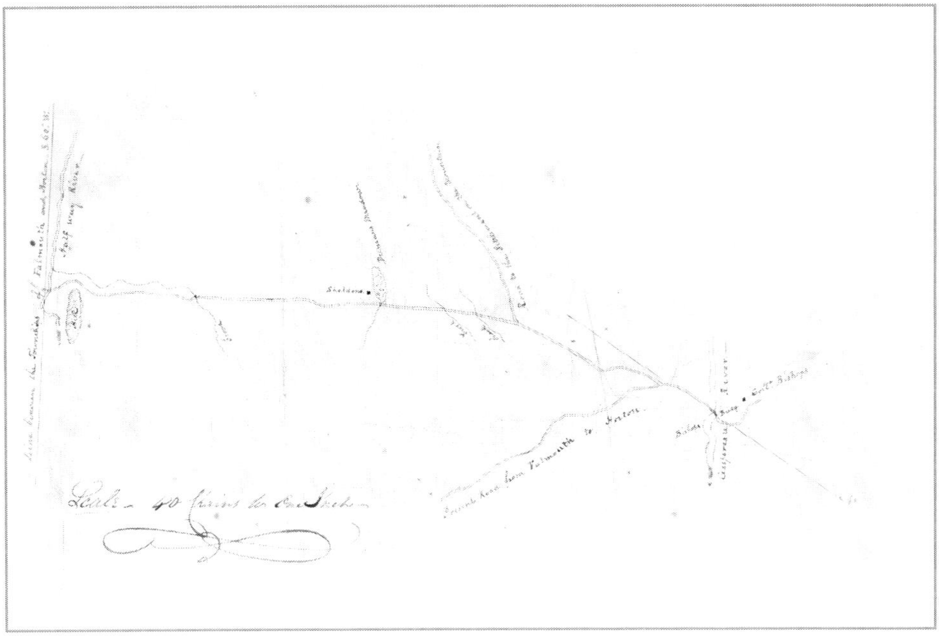

John Shey. Section of "Plan of the Road from Three Mile Plain near Windsor...to the New Bridge on the River Avon & thence to Bishop's Bridge," 1800. (NSARM 4.1.2.1) The road over Horton Mountain cut a considerable distance off the route from Halifax to Horton.

covered bridge, as illustrated, crossing the Avon from Water Street as Trunk 1 does today. While traffic from Halifax or the South Shore might still take the Upper Falmouth route, much of the area's business required a direct link between the Annapolis Valley and the Town of Windsor, so the Mount Denson Road received more attention than before. As a result, and in spite of attempts to make it more user-friendly, the hilly road from Upper Falmouth gave way with time to the lower route through Mount Denson, which evolved into Route 1 with some variations.

Despite many gradual alterations to this Annapolis Valley route, portions of former roads are recognizable today. The direct road over the hills to the Gaspereau River has been abandoned by through traffic for many years, but it can still be traced as a road leading straight ahead from the bridge over the Avon; it quickly deteriorates into a track over the Grey Mountain but becomes a paved road again on the far side. A section of the Post Road between Grand Pré and Wolfville still exists and is identified at Grand Pré, where it runs in a straight line over a hill. An improvement was made by diverting the road with a curve to the south—the route now followed by Trunk 1—thus avoiding the hill. More recently, both these

William Eagar. "Windsor, N.S., From Fort Hill," 1840. The covered bridge spans the Avon River to the right in this sketch.

routes have been superseded as the "main road" by Highway 101, which runs even further south, parallel to the two earlier roads.

The Old Annapolis Road to Halifax, 1815

We likely have either heard or said, "It seemed like a good idea at the time." So did the late eighteenth-century project to provide a direct route between the old and the new capitals of Nova Scotia. Surveying for a road from Annapolis to Halifax had been started as early as 1776. This ambitious project was backed by Governor Parr, and in 1784–85 John Harris surveyed a line through completely uninhabited country from Annapolis to Halifax. The aim was to provide a shorter route and to settle people in the province's interior. Construction was started at the western end of the line, and a few settlers received grants in this area. After only a few miles had been developed, the project faltered, until 1815 when Harris resumed his work on it. His map of that date shows a road that ran south of his earlier line, and which, as he explained in his notes on the map, took into greater account the resources that would encourage settlement. "By climbing trees and exploring to right & left, it was often discovered that good hills of hardwood lay at no great distance." The eastern portion of the road was intended to link up with the Hammonds Plains Road out of Halifax, but the phase east of Sherbrooke (today New Ross) presented such difficult terrain that the project was stalled. Work was begun from the Halifax end, where a route was constructed westwards from the Hammonds Plains Road towards Card Lake, but only fragments have survived.

The Old Annapolis Road did not become the major route that Governor Parr had expected, as the eastern section was never successfully completed. Between Sherbrooke and Annapolis, some settlers received land grants and were eventually established. Among these settlers were groups of disbanded soldiers who had fought in the War of 1812. Captain William Ross and his followers received land at Sherbrooke in 1816; others received grants in what was known as the Dalhousie Settlement in recognition of the lieutenant-governor's advocacy for its establishment.

The road was cut out and work began in 1816. Harris was responsible not only for determining the line of the road but also for laying out the lots for would-be settlers. The notes on his surveys suggest that he took a personal interest in these settlements. In 1818 he surveyed "The Second Location of Land in the Settlement of Dalhousie," at the crossing of the road from Nictaux to Liverpool (only partially extant today) and "the new road called Harris's Road from Annapolis to Halifax via Hammonds Plains," today known as Albany Cross. Harris numbered the lots and classified them: "such as by cultivation may be made good farms," those "of middling quality," and those which appeared to be "of so bad a quality as not to admit of cultivation."

Only a few people at first had sought property in the area, and not all had taken steps to develop it. Near Musquash and Duck lakes (today known as Spectacle Lakes), William Fairn already owned two hundred acres. Horatio Nelson Harris owned the adjoining lot of one hundred acres, and Ben Jay Fairn had two hundred acres around Paradise Lake. Harris's survey superseded an earlier plan in which lots had been laid out on a different system but most had not been granted. Harris observed that some had been "improved" (i.e., some clearing and/or cultivation had occurred), and he recommended that the people who had worked on them should have a chance to take up the newly surveyed

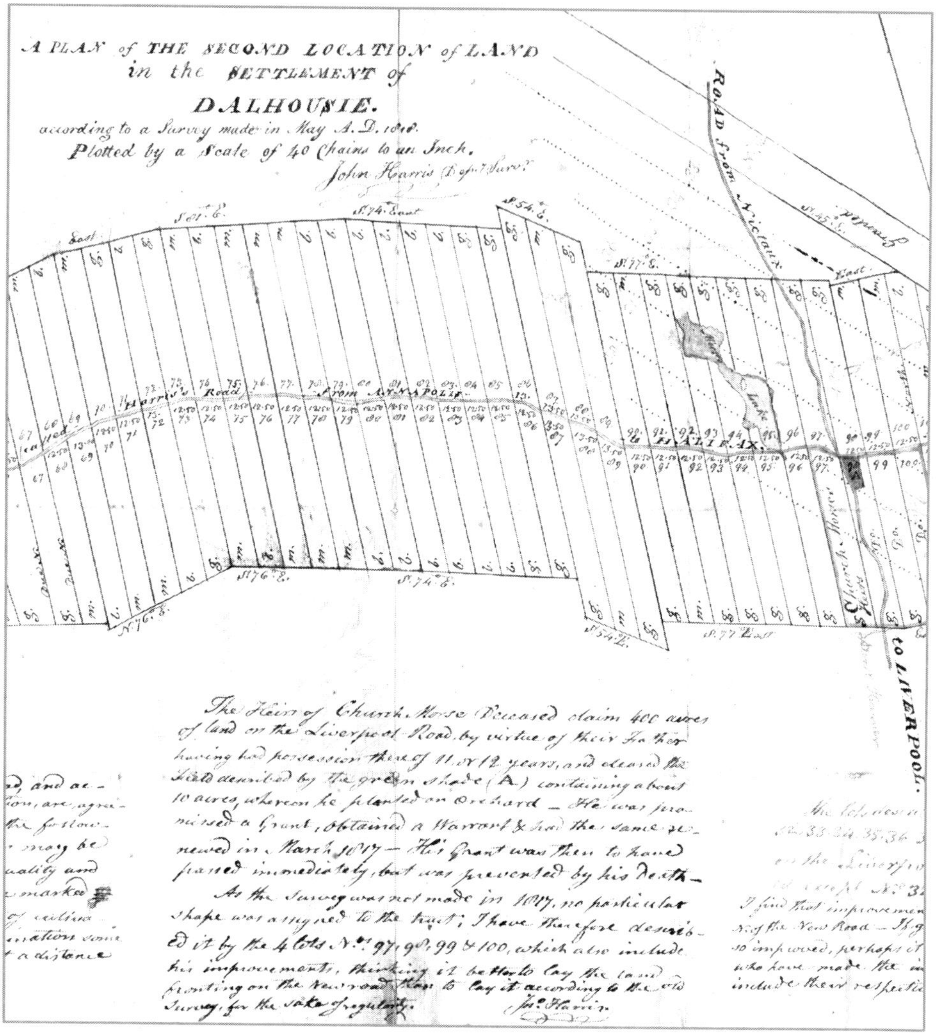

John Harris. Section of "Second Location of Land in the Settlement of Dalhousie," 1818. (NSARM RG20 Series C Vol. 85 No 152) This portion of the map shows the new road, the lots Harris had laid out for settlement, and the property claimed by Church Morse's heirs.

land. The only identified grantees at the crossroads were John Gall and the heirs of Church Morse. Morse's land had been the object of a survey by Harris two years earlier, in which he recounted the story of the misfortunes of this settler, "an industrious man" with six children, for whom he was endeavouring to obtain a grant. By the time the 1818 survey was made, Morse had cleared about ten acres at the southeast corner of the Liverpool Road and planted an orchard. "He was promised a Grant, obtained a Warrant & had the same renewed in March 1817. His Grant was then to have passed immediately, but was prevented by his Death." Harris claimed the land on behalf of Morse's heirs, but I have failed to find any evidence that he was successful.

Lord Dalhousie visited the area in October 1819, about two years after parts of the road had been constructed. It was a rainy day, and he was not impressed with the conditions: "From [the crossroad with the Liverpool road for] twelve miles...the road is bad—seven miles of it a swamp, but it appears to me a dry line might have been found, as I observed the land swelling and sloping on both sides of us....From Dargie's depot to Annapolis ten mile[s] very bad road, tho' easy to mend & much public money has already been voted for it." Altogether, he covered about twenty-seven miles of the new road, but he did not visit the section between the Liverpool Road and Sherbrooke at that time.

Only parts of the route laid out by Harris still exist either as sections of the present road from Annapolis through Lake LaRose, the Dalhousies, Forties, and New Ross, or as logging roads and trails. A short section of Trunk 10 follows Harris's route for a few miles east of Albany Cross. East of New Ross, a different road was eventually established in the 1860s, taking a route north of Harris's proposed line between Sherbrooke and the Windsor Road. Small sections of the Old Annapolis Road can be found today near Card Lake where the Halifax end of the road resumed; at Rafter Lake where Bowater Mersey has established a hiking trail; and at Wrights Lake, near Upper Hammonds Plains, from where the road runs to join the main Hammonds Plains Road. These fragments are all that remain of the eastern end of Governor Parr's projected road.

The failure of the Old Annapolis Road to become a viable highway left the South Shore and Annapolis Valley routes to be developed as the main routes to southwestern Nova Scotia. They evolved, with some changes over time, first as Trunks 1 and 3 respectively. Now, Highways 101 and 103 have mostly superseded the older roads.

Michael Seymour. From the Preston Road, near Dartmouth, 1845.
Travellers to Eastern Nova Scotia could take this road from the Dartmouth ferry to Preston on the first stage of their journey.

5

Eastern Nova Scotia

Roads in eastern Nova Scotia at the beginning of the nineteenth century were primitive. A map made in the early 1800s shows no major roads beyond Pictou and nothing at all along the Eastern Shore. As the century progressed and settlement expanded, the roads improved. The Post Road continued through Antigonish and Tracadie to the Strait of Canso, and a road ran from Tracadie to Guysborough. An important route linked Antigonish with Sherbrooke to the south by way of the West River of Antigonish and St. Mary's River, and another road followed the South River to St. Andrews, Country Harbour, and Guysborough.

Along what we know today as the Eastern Shore, a short land route had in fact existed, linking Halifax and Lawrencetown from the time of the latter settlement's establishment in 1754. However, the road farther eastward, to Sherbrooke and Canso along what is today known as the Marine Drive, was slow to develop. It would be some time before a continuous road linked the coastal communities. Further to the north, a road was established across the spine of the province between Halifax and Guysborough by way of Musquodoboit. As in other parts of Nova Scotia, many early roads were narrow, hilly, and unsuitable for wheeled traffic, and these were either altered or abandoned in favour of more satisfactory routes.

The Governor's Road

In 1792 an attempt had been made to create a direct road between Halifax and Pictou by way of Dartmouth (with a ferry crossing from Halifax), Preston, and Musquodoboit. Governor Wentworth personally launched the undertaking. His arrival in Pictou in September of that year is described by George Patterson in his *History of the County of Pictou.* Patterson was surprised to find that by the 1870s, the road already seemed to have been forgotten. The governor came to Pictou by sea, bringing with him "provisions, axes and hoes, bought with the proceeds of a Government lottery" for the five hundred men who had agreed to work on the road. He then proceeded to ride along the unfinished route to Musquodoboit. It was known as the Governor's Road and had an auspicious beginning with a celebratory ball at Government House. Although parts of it survived into the early 1800s, it did not replace the route through Truro as the main road to Pictou. By 1818 the southern section was already disused. John Chamberlain's map of that date shows "the Old Wentworth Road newly abandoned," branching off at Preston from the road out of Dartmouth towards the north end of Porters Lake and angling up towards Elderbank along East Brook. A dirt road that appears to follow the same route, at least in part, is identified in a map issued by the Geological Survey in 1909 as Lookout Hill Road. A trail still runs along East Brook from the head of Porters Lake.

The Governor's Road was replaced by one that followed the lower part of the Musquodoboit River "by way of Gibraltar" towards Meaghers Grant. It reached Pictou along the west branch of the East River to Pictou Harbour, where a final ferry ride completed the journey. For years this road was considered a possible alternative to the Truro route to Pictou. The 1820s Belcher's *Farmers' Almanack* listed it as such, with the names of inns along the way. This route fared no better than its predecessor as a main road to Pictou and was dropped from the *Almanack* in 1842.

The Road through Pictou to the Strait of Canso

The most convenient route from Halifax to Pictou in the early nineteenth century remained, as it is today, by way of the already established road to Truro. The journey from Truro to Pictou had initially been rather a hazardous undertaking. Richard Uniacke recalled that the road had been non-existent in 1775, when he and some companions travelled on foot through woods without even a blazed trail to follow. In 1786, when the Reverend James MacGregor first arrived in Pictou from Halifax, there was a trail of sorts from Truro fit only for travel on horseback and even that was difficult.

By 1816 the road had so improved that the stagecoach service established by Ezra Whitter from Halifax to Truro was continued by Jacob Lynds as far as Pictou. The following year Lord Dalhousie described it as a delightful ride. The onward journey from Pictou to Antigonish, however, was less inviting, and the lieutenant-governor proceeded by water to Arisaig before continuing his travels on horseback. In 1829 the Eastern Stage Coach Company was formed, combining Witter's and Lynds's routes between Halifax and Pictou. By 1833 the road was good enough to allow the service to extend to Antigonish.

The main road into Pictou Town, later known as the Old Mount Thom Road, ran over Mount Thom and along the West River. Surveyor Peter Crerar improved this road in the late 1820s. Joseph Howe, travelling on horseback in 1829, wrote

John Elliott Woolford. "Crossing the West River of Pictou towards Merigomish," 1817. (Nova Scotia Museum, 78.45.93) The road eastwards from Pictou crossed this precarious-looking bridge.

of "struggling up the side of Mount Thom," but observed that the alteration in the road that was almost completed "[would], in future, be able to avoid the tiresome ascent and descent." Howe wrote of a man named Stewart, who had been the first mail carrier on the eastern road in about 1805 and now lived in retirement at the top of Mount Thom. In the early days, Stewart had made the journey to Halifax on foot, parts of the road being unfit even for a horse. He used to carry the mail in his pocket and "a small wallet of oatmeal" over his shoulder to sustain him on the journey. A quarter of a century later, he could watch the Eastern Mail passing with a horse and a gig that was so heavily loaded the horse struggled to ascend the hill. Howe declared the road from Mount Thom to the West River "rather monotonous and tiresome" and was glad to reach Archibald's Inn on the West River Road and change horses for the final ride into Pictou.

The early nineteenth-century Post Road eastwards from Pictou was very different from today's Trans-Canada Highway. Between the West and Middle rivers,

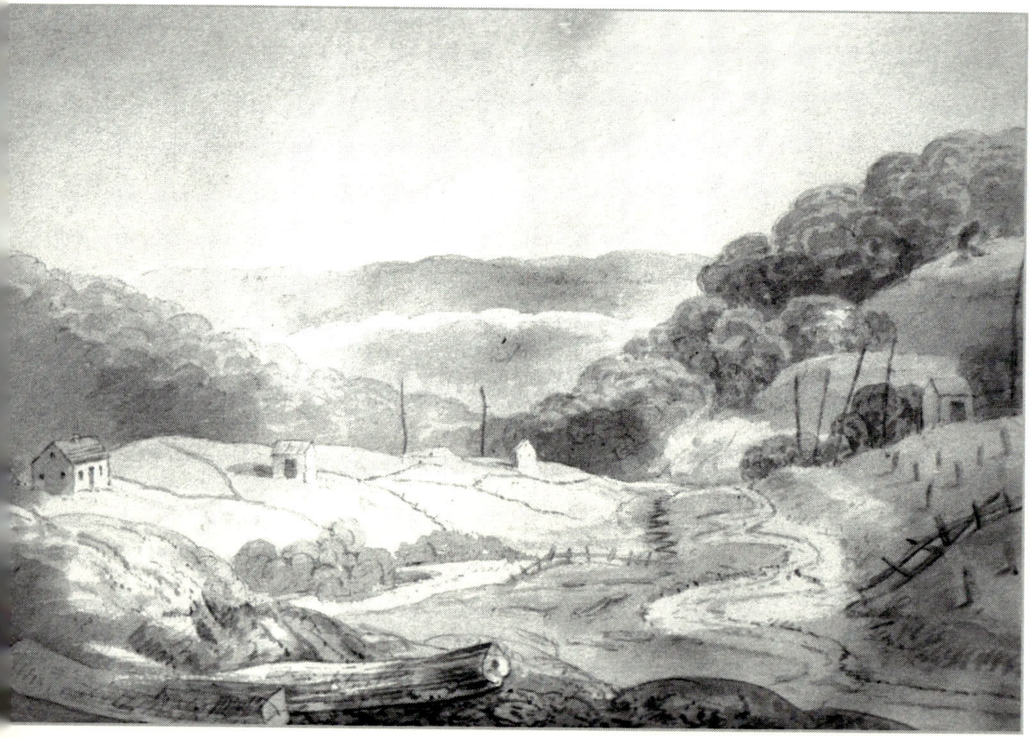

John Elliott Woolford. "On the Green Hill Between the W. & Middle River of Pictou," 1817. (Nova Scotia Museum, 78.45.94) The ascent from the bridge up Green Hill was a challenge even for mail horses.

it passed high over Green Hill, which was later avoided by a loop to the south. It clung to the coast as it headed towards Antigonish and was replaced as the main road by a straighter route in the 1840s. Like the other roads in the province, it underwent changes as traffic increased. Roads running southwards linked it to Sherbrooke, Country Harbour, and Guysborough. From Antigonish to the Strait of Canso, it ran through a number of small communities populated chiefly by Highland Scots and Acadians, and also passed through a Mi'kmaq reserve. At Aulds Cove, a ferry across the strait provided a link with Cape Breton Island. In the twentieth century this road became Trunk 4. Today it has been replaced by Highway 104 as the main road to the Strait of Canso. The ferry has been replaced by the Canso Causeway. Local traffic still uses much of Trunk 4, though parts have become integrated into Highway 104. Traces of even older roads can still be found, so that in some areas three generations of highway exist side by side.

The Changing Road: Antigonish to Tracadie, ca. 1847

In 1846–47 work was undertaken to replace the sometimes circuitous route of the old Post Road between Antigonish and Tracadie. Two interesting maps by James McKenzie, a draftsman with the Crown Lands Office who compiled maps of many parts of Nova Scotia, chart these changes. One represents the road from Antigonish to just west of Pomquet, and the other shows the road eastwards from Pomquet to Tracadie Harbour. They show both the original Post Road and changes that were planned to shorten and improve the route. As well, they give an interesting profile of the population of that area. McKenzie was one of a family of surveyors and draftsmen who worked in different parts of mainland Nova Scotia and in Cape Breton during the middle of the nineteenth century.

"Plan No. 1," from just east of Antigonish to just west of Pomquet, records that this part of the new route was laid out by John G. McKenzie in 1846, and the road itself was partially opened by Adam and Valentine McKenzie that same year. A note on the back of the second plan, which shows the longer stretch east of Pomquet to Tracadie Harbour, tells us that improvements to this section of the road had been "approved and confirmed" by council on December 4, 1846. An interesting feature of the maps is that they list the landowners over whose property the new line of road passed with the distances involved for the purpose of providing compensation.

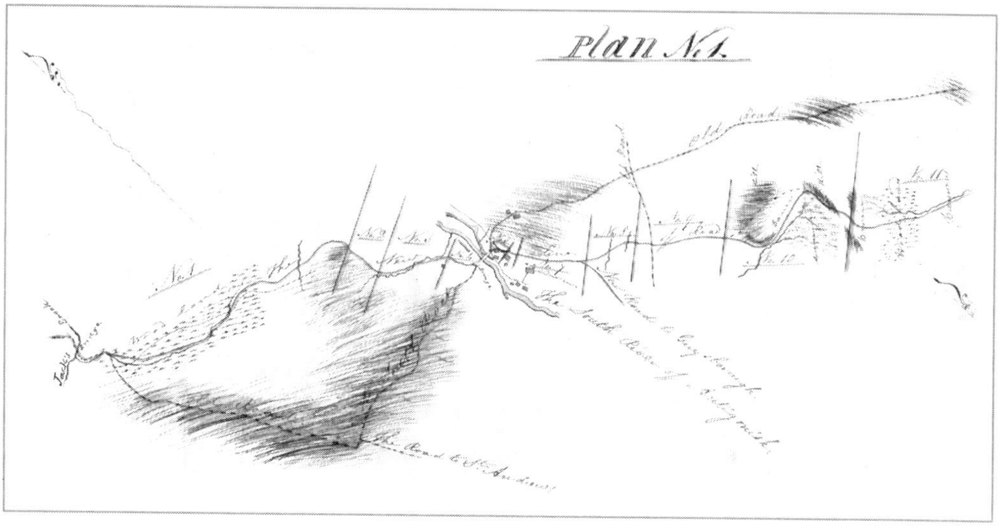

James McKenzie. Part of "Plan of two sections of Road one on each side of the South River of Antigonish, 1847." (NSARM *4.1.1.10) In this portion of the map we see the older roads and the recently surveyed new road.*

Before the improvements of the 1840s, the main road eastwards from Antigonish turned inland from Jacks Brook, just east of the town, then formed a right angle back to the bridge across the South River of Antigonish. A road still runs south from Exit 33 to Beech Hill, but today only a power line and a trail mark the former route to the South River Bridge. McKenzie's new line lay closer to the shore following a branch of the brook, curving around a hill, and following another brook before reaching the bridge. Much of this route has since been absorbed by Highway 104.

From the South River, the new road ran south of the earlier line towards the head of Pomquet Harbour, again following brooks and valleys to avoid the intervening hills. A "new road to Guysborough" branched off to the right just east of the South River. The newer road followed a fairly direct route, except for one spot where it made a detour over the lower slopes of a hill to avoid a marshy area. It then ran through woodland and barrens to join a section that had been recently opened west of Pomquet.

The second map picks up the road east of Pomquet, where several route changes were planned. Before McKenzie's improvements, the Old Post Road turned northwards to follow the shore and survives today as a minor road through Church Cove. Another old road, known as the "Old Back Road," ran in a direct line across country with little regard for the terrain over which it passed. It rejoined the Post Road near the bridge over the Tracadie River. Highway 104 follows its line for a few miles east of Heatherton.

McKenzie's road was to replace the meandering shore route as far as Tracadie, where it merged with the Post Road for a short distance through the settlement. From there, it was the new route that clung to the shore, whereas the Post Road ran straight to the Tracadie River. Here McKenzie planned to establish a new bridge a little upstream from the old one.

The road underwent further alterations as Trunk 4 was developed, but today both roads have been superseded by Highway 104, which still runs over much the same route as McKenzie's road almost as far as Heatherton. Here it takes up the line of the Old Back Road for a few miles, while Trunk 4 still follows the old route from Exit 36A to Tracadie. The Old Back Road reappears today as a country road south of the 104, from near Afton almost to Exit 37, where it is absorbed again by the modern highway near Tracadie Harbour. The former Post Road along the shore immediately east of Pomquet Harbour is today only a minor road linking coastal villages. In the Big Tracadie area, where McKenzie replaced the more direct road by one that ran closer to Tracadie Harbour, Trunk 4 still follows his route for the most part but takes the original path through the settlement of Big

McKenzie's road still runs close to the harbour near Big Tracadie, northwest of Exit 37 from Highway 104. J. Dawson photo.

Tracadie. A brief section of McKenzie's shore route still exists as a minor unpaved road bypassing the settlement. A parallel road further west, identified today as the Afton Loop, reminds us that continued upgrading changed the line of this and many other roads.

The property owners McKenzie recorded on the maps provide an interesting glimpse into the mixed ethnic origins of this area's population in the mid-nineteenth century. They were mostly farmers, with a handful of fishermen and essential tradesmen—blacksmiths and millers. At the western end, close to Antigonish, the road passed through land owned by a group of Scottish immigrants—chiefly the Frasers and the Chisholms. One of their neighbours was the "Widow McKenzie." Hugh McDonald, Esq., the only proprietor graced with any form of title on this map, was a merchant and the local Justice of the Peace. He later became a member of the legislative assembly. Along with other entrepreneurial Scots, he had acquired considerable land holdings in this area, including his estate near the bridge over the South River of Antigonish. Further east, beyond Pomquet, this Scottish enclave gave way to a piece of land identified as the property of "French Acadians," who are not individually named. From here on, the population becomes more ethnically mixed, with Scottish and Irish settlers scattered among Acadian and Mi'kmaq families. "Peter Batist an Indian" was the nominal proprietor of several dwellings on Little River (Afton River today) marked as "Indian huts" and "Indian buildings." This was traditional Mi'kmaq territory, and parts of the area had been designated by the government since 1801 as the Pomquet-Afton reserve.

The Acadians in this area, some with Mi'kmaq blood, had returned here from exile in about 1774 and had received grants of land a few years later. They were joined by other French-speaking families of varying origins. One of the first Acadians to settle here was Pierre Benoit. He had lived in the Merligueche (Lunenburg) area in 1749, when he had acted as a pilot for Edward Cornwallis, who had been sailing towards Chebuctou Harbour. His descendant, identified by McKenzie as Simon Benva, was by then living on the Pomquet-Tracadie Road.

Among other Acadians were two men identified on the map as "Captain Joseph Jewara" and "Luke Jewara," whose surname was strangely anglicized and who were in fact members of the Gerroir family. Their ancestor, Jean Joseph Gerroir, had overcome the captain and crew of the vessel on which he and other Acadians were deported and escaped exile by going into hiding. He later settled at Tracadie, where his descendants, Joseph and Luke (Luc), lived in the 1840s. Captain Joseph Gerroir was one of the well-established members of the community; his land was

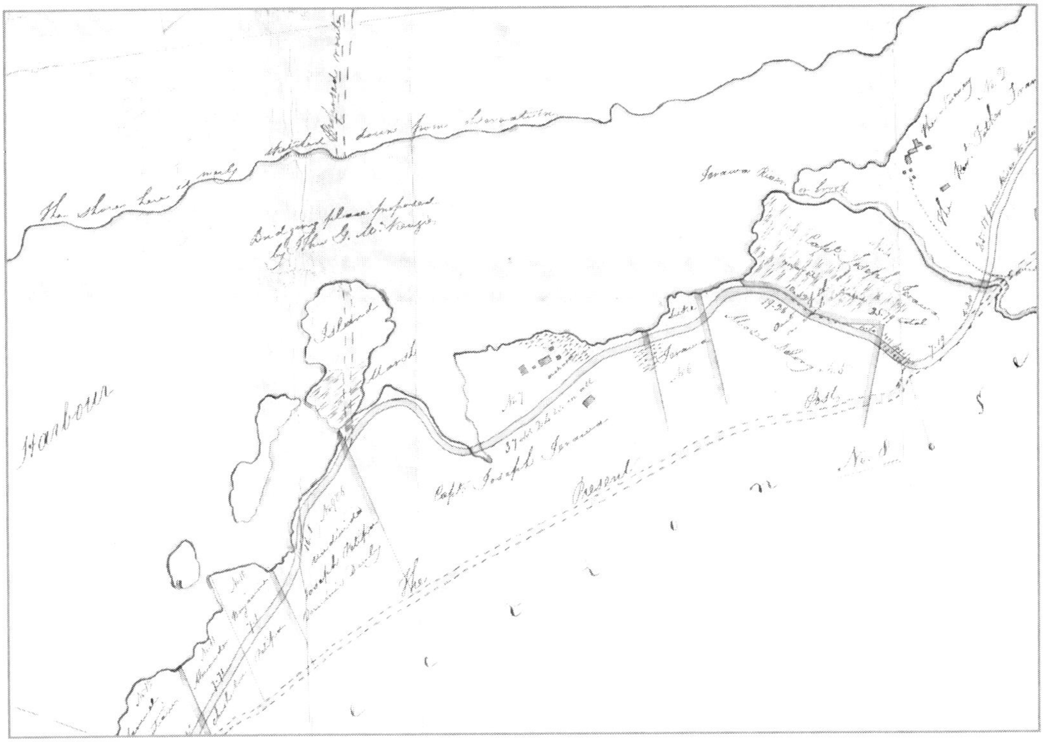

Section eight of James McKenzie's "Plan of the New Line of Road laid off in 1846 from the Pompket River to Tracadie River." (NSARM R/209 – 1846) The alteration followed the shore of Tracadie Harbour and passed through the Gerroir property before it rejoined the original road.

developed with an orchard and several buildings on his main lot. His smaller lot consisted chiefly of marshland and was bounded by a brook that separated it from the neighbouring church property.

Mingled among the Acadians and the Scots were a number of Irish settlers. This entire group, whose origins suggest they were mostly, if not all, Roman Catholic, were served by resident priest Father William Francis, who lived at what is today Big Tracadie, at the eastern end of the road shown on the map of section nine. Near the priest's house was a convent of Trappist nuns, founded in 1821, where a school was established for Native girls. A trail to the southeast led to the Trappist monastery of Petit Clairvaux, established in 1825, from which today's settlement of Monastery derived its name. The chapel that served the lay population stood closer to the centre of the community, on a point of land west of Tracadie Harbour. At an intersection known as "the Cross," a road gave access both to the chapel to

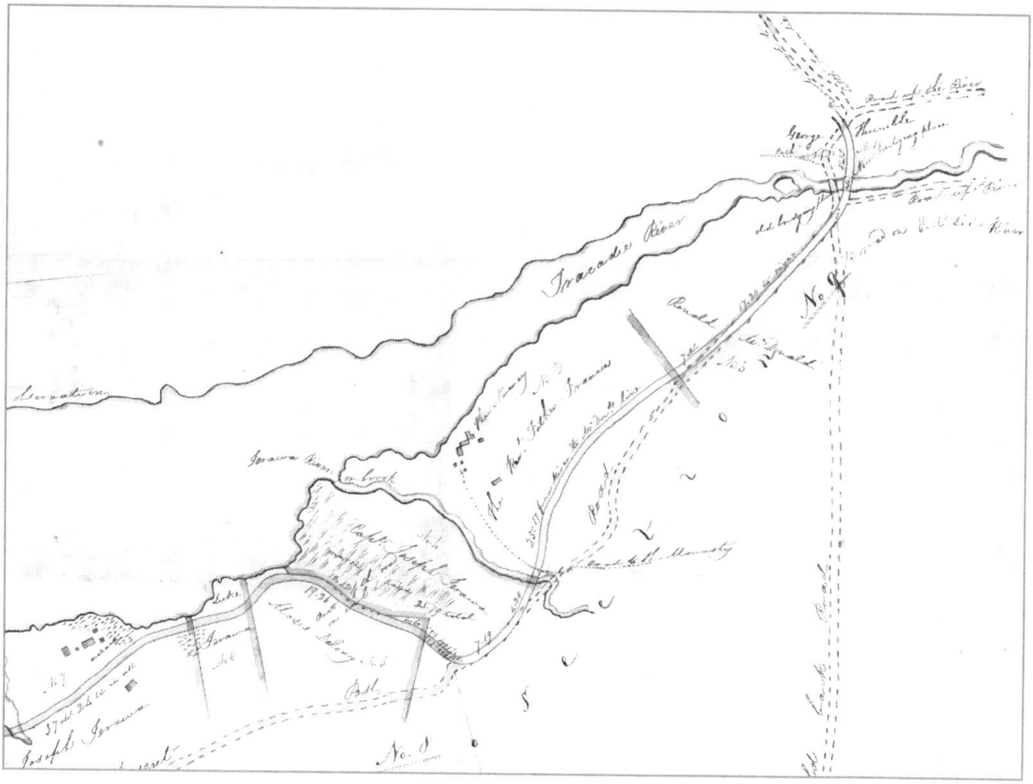

Section nine of McKenzie's "Plan of the New Line of Road..." (NSARM). Both the old and new roads passed through the church property and the McDonald land. The Old Back Road joins them just before the new bridge across the Tracadie River.

the north and the "Back Lands" to the south. Among the houses at the intersection was the Temperance Inn. As the only such establishment in this area, it must have done a good trade with passing travellers.

The ethnic diversity was further enhanced by a community of Black Loyalists on the Tracadie River. They had come to Nova Scotia with other Loyalist settlers and went first to Port Mouton and from there to Guysborough, but were unable to obtain land of their own. So led by Thomas Brownsprigg, they came to the Tracadie area, where they petitioned the government for land in 1787. In September of that year seventy-four Black Loyalists were granted three thousand acres in this area.

The people who used this road in the first half of the nineteenth century represented an interestingly varied population. Many of their descendants still live in the area. The modern highway bypasses these settlements, but travellers with time to spare can still trace parts of the road laid out by the McKenzies in 1846–47.

The Roads to Guysborough

The eastern end of Nova Scotia has always been relatively isolated from the rest of the province. Canso and Guysborough, former fishing stations under both the French and the English, remained accessible mainly by sea for many years. The long river estuaries on the Eastern Shore, such as that at Country Harbour, impeded the development of a satisfactory coastal land route to the far end of the province. In the early days of road building, a direct inland road from the capital to Guysborough was thought to be the best solution. A route was established from the Truro Road north of Halifax to the Musquodoboit Valley and the West Branch of St. Mary's River, then directly eastwards in a straight line over hilly terrain to Guysborough. Part of the western section of this route has survived as the Old Guysborough Road, which leaves Highway 102 near Stanfield International Airport as Route 212 and becomes Route 357 at Elderbank and then 224 from Middle to Upper Musquodoboit. Roads were also constructed through Preston to the Musquodoboit Valley. From the 1820s to the 1840s, improvements were proposed to this route, some of which were implemented, but it never became the major road eastwards from the capital that its proponents envisaged. Travellers to Guysborough more often went by way of Pictou or Antigonish, where routes ran southwards along river valleys, or by a route southwards from Tracadie Harbour through Monastery, which became today's Route 16.

The St. Mary's River Route, ca. 1830

In 1830 Joseph Howe left Pictou for Guysborough via ferry across the harbour and travelled by what he described as "eight miles of tolerable road" up the East River to New Glasgow and Stellarton. From New Glasgow, the main road ran eastwards to the shore near Sutherlands River and along the coast to French River, where Howe found the bridge across the fifty-yard expanse of water to be "in such a state of dilapidation that you can neither pass it on foot or on horseback." Travellers at that time had to take a scow ferry across the river in order to reach Copeland's Inn on the far side. The Post Road ran for several miles through the scattered agricultural settlement of Merigomish and continued along the

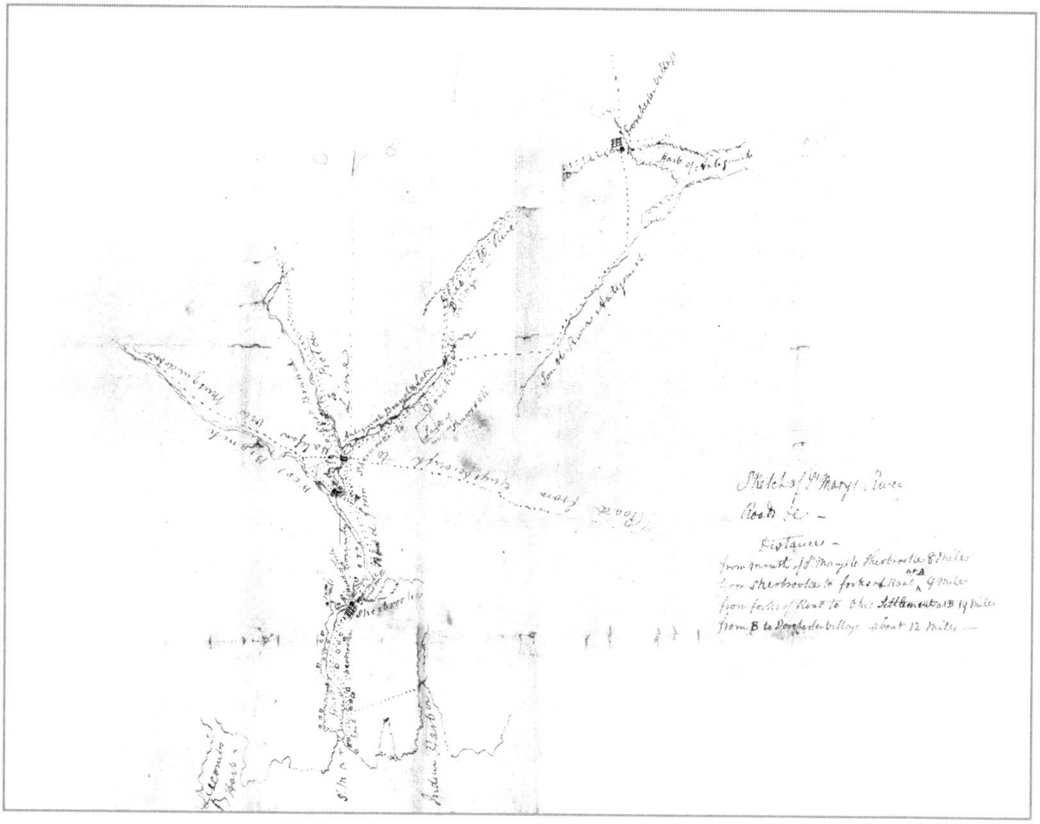

Anon. Section of "Sketch of St. Mary's River, Roads & c." (NSARM F/209 undated) Dorchester was the early name of the settlement on Antigonish Harbour. Several roads converged just north of Sherbrooke.

shore through Lower Barneys River. The stretch of road between the mouth of Bailey's Brook (Lismore) and Arisaig appeared to Howe to be "the finest part of the Gulf Shore settlement," though he considered the inhabitants to be unduly frugal in their way of life. Even the more prosperous ones "instead of supplanting their old log houses with buildings of a more roomy and comfortable character... still reside, with families of ten or a dozen children, in the same shielings...as when their infant clearing was commenced." From Copeland's Inn to Antigonish the road was devoid of what he described as "decent Houses of Entertainment." A few miles after Arisaig, the Post Road turned off sharply to the right to run across country to Antigonish, where hospitality was at last provided by Mrs. Simmons's Hotel.

Howe also described a shorter route known as the Mountain Road that ran inland directly to Antigonish. It reduced the journey by ten miles but was passable only on horseback, being too narrow for a carriage. This route would eventually be improved and supplant the road round the shore as the main road eastwards. Today, Route 245 still follows the shore. Trunk 4 takes the more direct route to Antigonish. Highway 104 bypasses the smaller communities and in some places has absorbed Trunk 4 and the older Mountain Road.

Several roads from the Gulf Shore to the Eastern Shore and Guysborough followed natural routes along river valleys. The road from Pictou to Sherbrooke ran along the East Branch of St. Mary's River and joined the Guysborough road near its confluence with the West Branch. Two roads led from Antigonish to Sherbrooke, one by way of the West River of Antigonish and the Ohio River, the other along the South River. They all came together at the head of Lochaber Lake, and followed the east side of the lake and St. Mary's River to the settlement of Sherbrooke and the river mouth. They also intersected with the direct east-west road just north of Sherbrooke. An undated map shows these roads as they were in the early nineteenth century.

William Moorsom. Lochaber Lake, *from his* Letters from Nova Scotia, *1830 (NSARM). Moorsom had approached the lake from the south, by a narrow road that was clearly more suited for horses than for wheeled vehicles.*

We rejoin Joseph Howe as he leaves Antigonish on his journey to Guysborough. He chose the new road along the South River of Antigonish rather than the older one through the Ohio settlements because it was shorter. He later regretted this choice, writing that "the old road…is tolerably good, and winds along through some of the finest farms of the Upper District. The other is a new one, recently explored, wet and muddy enough in some places, and plentifully strewed with roots, which when they run along upon a soft surface, form natural traps for the legs of a horse." The first ten or twelve miles were dull and wearisome, but he was delighted with the view of Lochaber Lake, referring to the recently published sketch of the lake by William Moorsom, who had passed that way a year or two earlier.

There were a few settlers around the lake at this time, but a large tract of land that had been set aside as an endowment for King's College had not been developed. Howe found the road from the lake to the confluence of the branches of St. Mary's River uninteresting and "of indifferent quality." He passed a few farms and mills before coming to Archibald's Inn at the forks, near the intersection with the Halifax-Guysborough Road, and eventually arrived at Sherbrooke. This settlement he described as "a rough and unsightly cluster of wooden houses" at the head of navigation, whose inhabitants were chiefly occupied with sawmilling, shipbuilding, and the timber trade. Its gold rush glory days were yet to come.

Howe retraced his steps northwards to breakfast at Archibald's Inn before taking the road to Guysborough. Between St. Mary's River and the outskirts of Guysborough, the road at this time was in a bad state with few settlers. This was the original road following the direct hilly route shown in Wentworth Taylor's 1840 map (see page 108). Howe was not impressed by its scenery: "For many miles along this road the land is indifferent…old and abandoned clearings, broken fences and dilapidated log huts." The road was sparsely settled, and the only place offering refreshment was a small cottage about ten miles from the forks. After a long day's ride, Howe finally arrived at a "public" in Guysborough run by Christian Miller, the sheriff of Sydney County—the administrative area that was later divided to form Antigonish and Guysborough counties.

A Direct Route to Guysborough, ca. 1840

Wentworth Taylor, the surveyor for eastern Nova Scotia, began to work on an alternative route to the original Halifax-Guysborough Road as early as the 1820s. The western section of Taylor's planned new road shows that the "main road" ran north of the Musquodoboit River, as it does today, while his proposed new route ran further south. This proposed new road began twenty-nine miles out of Halifax, where the recently constructed road from Preston reached the Musquodoboit River near what is today Elderbank. From there, the proposed route ran south of the river to rejoin the main road close to where Henry's (today Chisholm's) Brook runs into the West Branch of St. Mary's River.

Parts of this parallel route can still be followed, passing through Higginsville (where several members of the Higgins family were settled on Higgins Brook in Taylor's day) and south of Mill Lake, near Upper Musquodoboit, where Route 224 (the Sheet Harbour Road) turns southeastwards. Taylor's route continues along the Pleasant Valley Road but quickly dwindles to a trail and is lost in today's Liscombe Game Sanctuary. The older "main road" to the north of the river has remained the chief route along the Musquodoboit Valley, continuing the Old Guysborough Road. It survives further eastwards as a minor road branching towards Trafalgar and West St. Mary's River.

The early method of surveying new roads—sighting by compass from hilltop to hilltop—resulted in roads that were direct but often extremely hilly. Such was the case with the original route from St. Mary's River to Guysborough, traversed by Joseph Howe ten years earlier. Another map made in 1840 by Wentworth Taylor showed the drawbacks of such a route. On this map, recording his "Exploration and survey of part of the Line for the New Eastern Road from Dartmouth to the Gut of Canso," he plotted a rather longer but far more practical route and compared it with the older one. The map shows both routes, indicating along the older road "hilly broken land," "high Hill," "great descent," "high Ridge," "deep hollow," "rapid descent," and similar hazards for wheeled vehicles. In explanation of his proposed new road, his annotations include "level round the hill," "easy descent," "level dry ground," "fine level for a road along the hillside," and "small ascent," which made for a more user-friendly highway.

Although the proposed route seems to have been constructed very much as he suggested, it did not become the major highway to Guysborough that Taylor had expected. Starting from Melrose on St. Mary's River, it survives as a minor road heading eastwards to Cross Roads Country Harbour. From there, it angles to the northeast, follows the Salmon River eastwards, and then heads northeast again towards Guysborough, very much as he planned it, avoiding the hilly area to the north. Sections of what appear to be the even earlier road also still survive in the Guysborough region. Among several other roads identified in

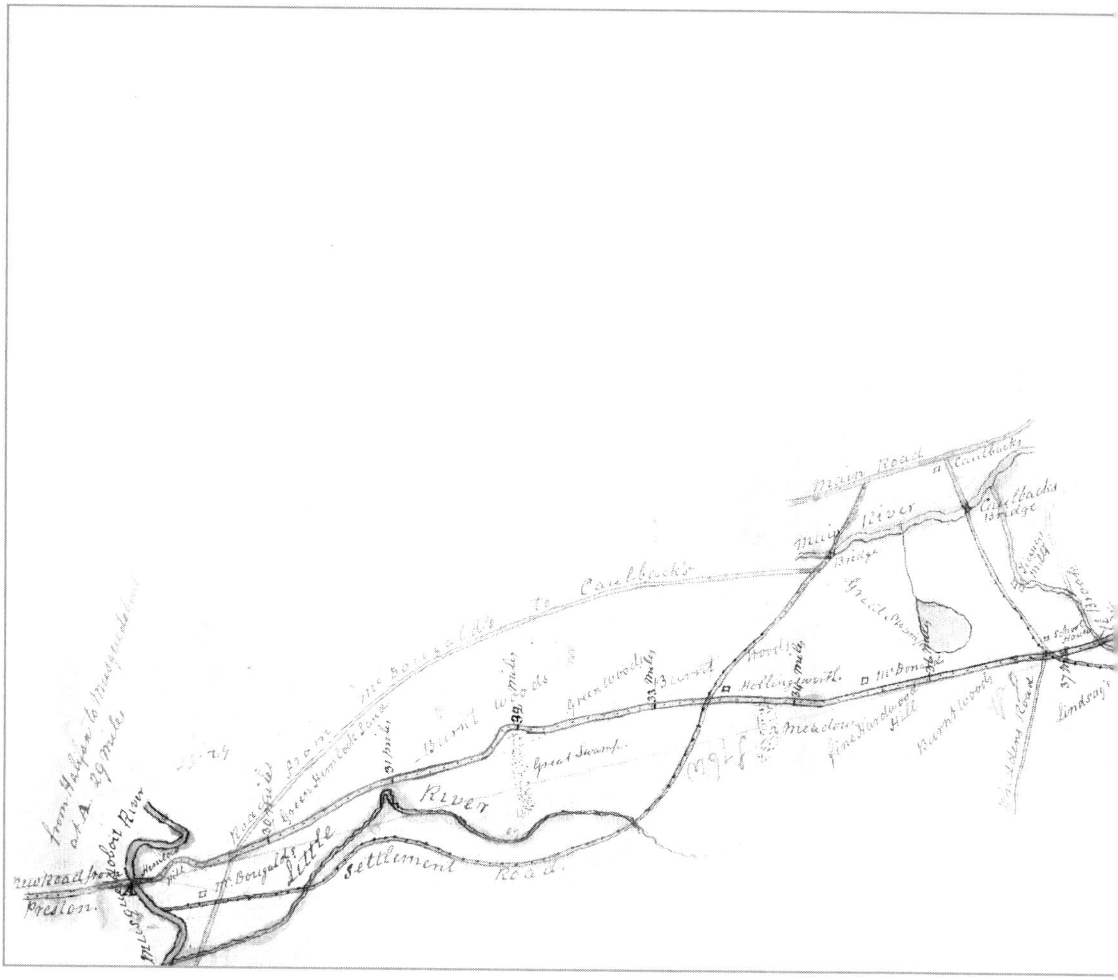

Taylor's plan was the "main road from Guysborough" that would join the road along the northern shore near Tracadie. Despite the efforts of Taylor and others to develop a direct route from Halifax, Route 16 remains the most popular road to Guysborough.

Much of the area the eastern section of the new road would serve was initially uninhabited, but pockets of settlement, particularly along the West Branch of St. Mary's River, and in the vicinity of Guysborough itself, appear on the map. Taylor records the names of the supposed owners of the land, without indicating whether they had actually cleared land and taken up residence. Some of them were unlikely to have done so. The proprietors on the West St. Mary's River included respectable members of Halifax society who would have considered

Western section of Wentworth Taylor. "Plan of a Survey of a Proposed New Line of Road from the West Branch of St. Mary's River to the Middle Settlement of Musquodoboit, to meet the New Road from Preston to Musquodoboit …" *[ca. 1820]. (NSARM F209/wm 1819) A small section of the "Main Road," today's Route 357, can be seen north of the river.*

their property an investment rather than a country retreat. Among the named absentee landowners were the Hon. H. H. Cogswell, a Halifax businessman who is unlikely to have spent much time on his property; Thomas Glencroft, Esq., about whom nothing more is known; John Tremain, Esq., (probably Jonathan, another businessman in Halifax and Dartmouth); and "the Heirs of the late Sir Rupert George, Baronet." Other more likely resident proprietors included William Archibald, William Smyth and his two sons, John McArthur, George McIntosh, and George Tates, who had acquired land originally granted to Donald McLeod and Hugh Ross.

The road runs through "the original St. Mary's grant," which was settled in 1800 by some members of the Taylor and McKeen families from Truro. Two

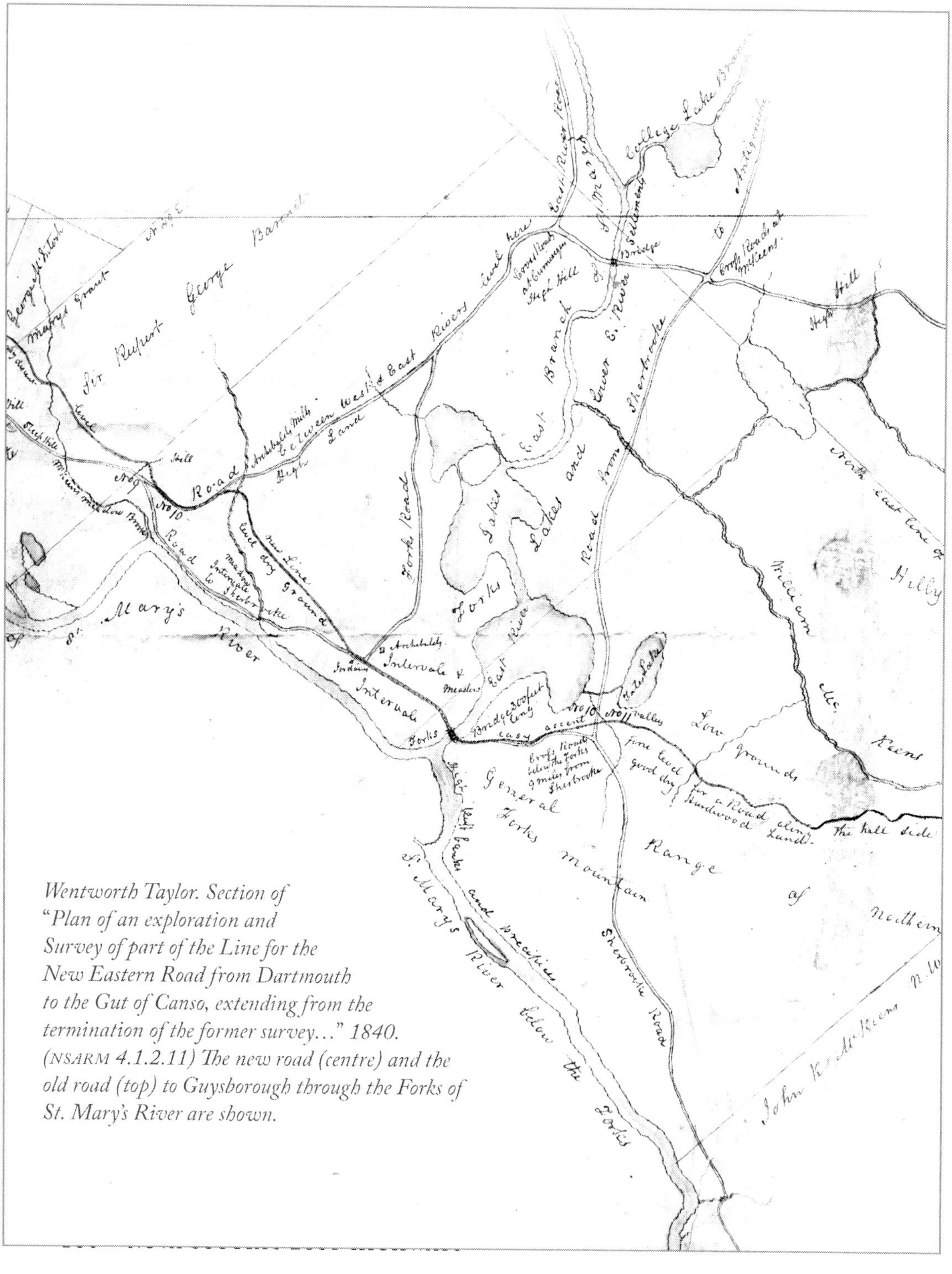

Wentworth Taylor. Section of "Plan of an exploration and Survey of part of the Line for the New Eastern Road from Dartmouth to the Gut of Canso, extending from the termination of the former survey…" 1840. (NSARM 4.1.2.11) The new road (centre) and the old road (top) to Guysborough through the Forks of St. Mary's River are shown.

other Taylor brothers came shortly afterwards, and John Taylor's son, Wentworth, became clerk of the St. Mary's Township, a Justice of the Peace, and the surveyor who planned this road.

An existing network of roads and bridges in the St. Mary's Forks area only a few miles north of Sherbrooke indicates a more advanced state of development here. The Archibald and Jordain houses are identified on the Forks Road, and Archibald's Mills stood on a stream flowing into St. Mary's River from the north. William Archibald, like the McKeen and Taylor families, came to the area from Truro shortly after his house was destroyed by a storm in 1813. He married and raised a large family on St. Mary's River. Peter Jordain, of French ancestry, moved from the Acadian settlement in Chezzetcook to the Forks area in 1812.

Taylor's road next passed through the sparsely populated Country Harbour Grant, where Loyalist settlers, including some blacks, arrived in the 1780s from disbanded regiments that had fought in the American Revolution. Approaching Guysborough, the road traversed land settled by an increasing population, but including, once again, land acquired by speculators such as Dartmouth business partners John Fairbanks and Robert Hartshorne. Lord Dalhousie had observed in 1817: "Mr Hartshorne...has acquired immense tracts of land, but of which he makes no use. I am told he is laying it off in lots of two hundred acres each which he sells at one pound per acre, thus obtaining £23,000 without laying out a shilling." There were also more genuine settlers, such as the Horton family who gave their name to the brook feeding into the Salmon River. The Dean and Chapter of the Anglican diocese of Nova Scotia held title to land south of the road, and closer to Guysborough there were a number of five-acre farm lots laid out on what was known as the Canada Settlement Road. Part of this road was to be absorbed in the planned "New Eastern Line" before it entered the town of Guysborough.

Today, some tourists headed toward Guysborough take the more leisurely route along the Marine Drive through Canso, while those in a hurry use the Trans-Canada Highway and Route 16. Few people today approach Guysborough over the remnants of Wentworth Taylor's road, but this road does still provide an opportunity to explore a relatively unknown area of Nova Scotia. Today's roads in eastern Nova Scotia are the result of the work of Taylor, the McKenzie family, and others who managed to establish practicable routes over what was sometimes inhospitable terrain. Along the Gulf Shore, this was a relatively easy task, and both Trunk 4 and Highway 104 represent further improvements over the original routes they absorbed or bypassed. The road to the Strait of Canso and Cape

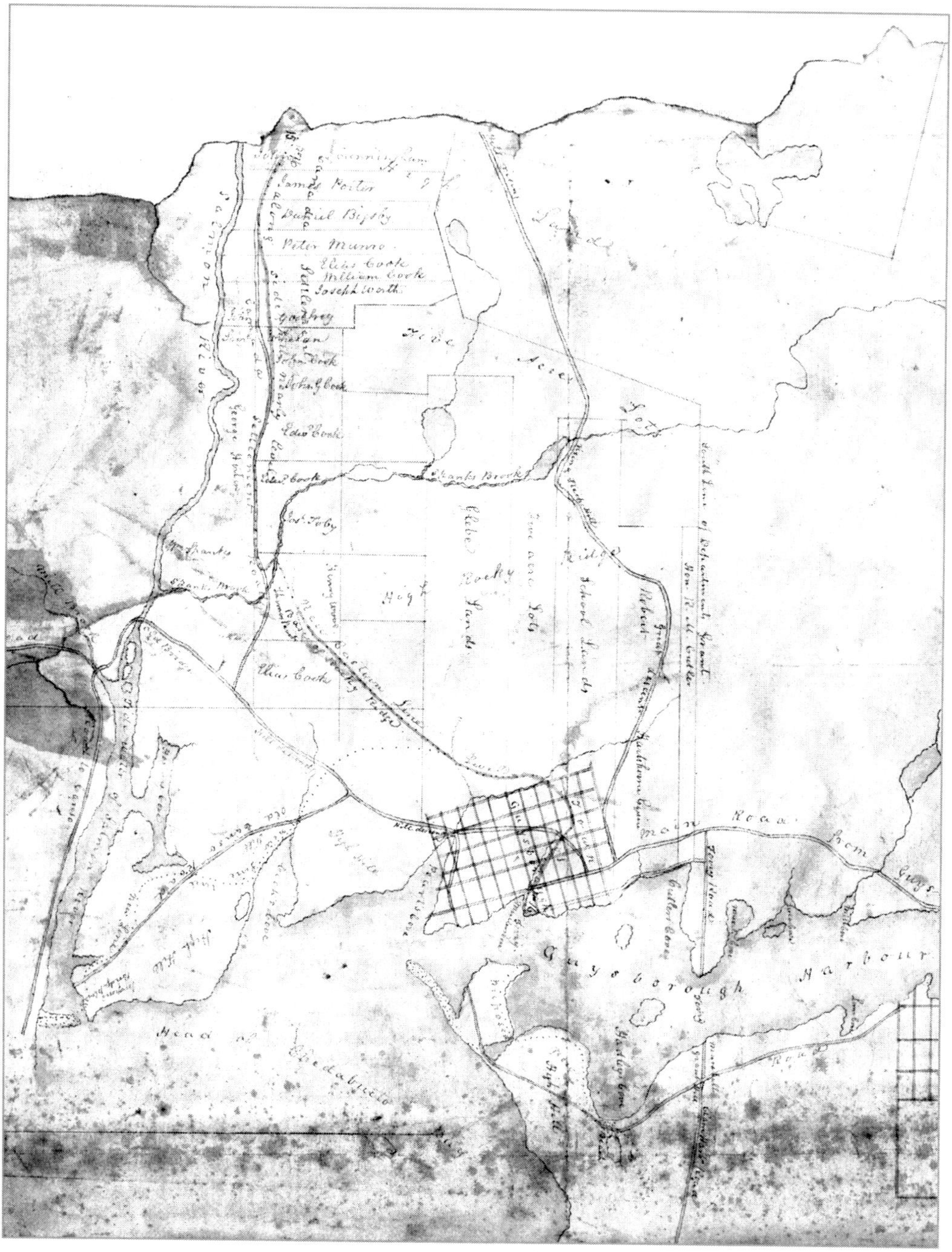

Eastern section of Taylor's map shows the approaches to Guysborough. (NSARM *4.1.2.11*)

Breton is now part of the Trans-Canada Highway. But the Halifax-Guysborough route presented greater challenges. The relatively small population in that part of the province has provided little incentive for major highway construction, so that many back roads still follow old paths that were planned as part of a main road that has not survived. Trunk 7 turns northward to Antigonish, and there is still no trunk road leading directly to Guysborough from the west.

6

Northern Nova Scotia & Cape Breton

ALTHOUGH NORTHERN NOVA SCOTIA PROVIDES the overland link between the peninsula and the rest of the country, road travel between Halifax and the New Brunswick border presented a problem for many years. The Cobequid Mountains were a formidable barrier between the town of Amherst and the rest of the province. In the early nineteenth century, many travellers from Halifax took the water route across the Minas Basin to Parrsboro and then the road northwards to Amherst. The longer overland route ran westwards from Onslow along the shore of Cobequid Bay then inland through Londonderry to River Philip and Amherst. The route eventually taken by Trunk 2 follows a combination of the two, as it runs along the shore to Parrsboro and turns northwards through Halfway River before swinging off through Springhill on its way to Amherst. For many years, Trunk 4 over Folly Mountain and through the Wentworth Valley was the more usual route. Modern engineering equipment has made the more direct route taken by Highway 104 over the Cobequid Pass possible.

Cape Breton was another difficult and inaccessible area. From the 1780s until 1820, it was a separate Crown colony with its own administration. It was sparsely populated, with centres of settlement in the Sydney area in the north and around the Inhabitants River and Isle Madame in the south. Overland communications between the two areas were primitive, and most travel was by water on the Bras

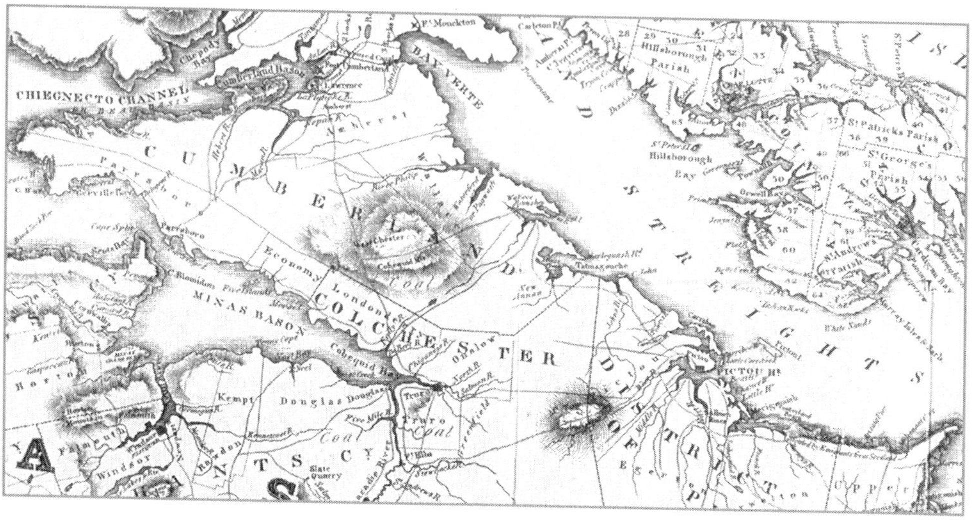

Part of William MacKay's New Map of Nova Scotia compiled from the latest Surveys for the Historical and Statistical Account of Nova Scotia, 1829 *showing the roads in Colchester and Cumberland counties.*

d'Or Lakes or along the coast. The Strait of Canso was also an obstacle for travellers. When Cape Breton was re-annexed to Nova Scotia in 1820, improvements to the road system were undertaken by surveyors from the mainland, and ferry operators received a government subsidy.

Roads in Colchester and Cumberland Counties

The early roads in northern Nova Scotia kept as much as possible to the river valleys. One of the oldest routes in this area of the province was a drove road established by the Acadians and identified in a map, probably made in the 1740s, as the "Portage from Cobequid to Tatamagouche by which cattle are transported, destined for Isle Royale" (author's translation). The road along the North River remained in use in the nineteenth century but lost the strategic importance of the pre-Expulsion days.

Because of the difficulty of traversing the Cobequid Mountains in travelling from Amherst to Truro, an easier route had been established in the eighteenth century by British settlers wishing to take their cattle to market in Halifax. They established a trail southwards from Amherst whereby they could drive their animals as far as Parrsboro. From there the livestock were carried by boat to Windsor to take the relatively easy road to the capital.

Of course, not only cattle had to travel through northern Nova Scotia. Mail and passengers needed access between the settlements in the northern part of the province and the capital, as well as to and from the rest of British North America. Although there were well-established shipping routes across the Minas Basin and the Bay of Fundy, land routes were also required. In the early nineteenth century a network of roads was developing in Colchester and Cumberland counties.

Travel to Amherst, 1818

In the early 1800s, a commercial passenger service across the Minas Basin ran from Windsor and Horton on one side to Parrsboro on the other from April until Christmas, two or three times a week as business warranted. A road led northwards from Parrsboro to Amherst, and many travellers used this convenient route to the border.

When Lord Dalhousie's party travelled to Amherst in 1818, they took this route through Parrsboro, which was by then well populated with Loyalist settlers. They started out from Horton, in "a very good sloop packet carrying one or two gigs and horses." The voyage was not without incident, but they arrived safely in Parrsboro, where a Mr. Ratchford offered hospitality at Partridge Island. The isolated settlement was, according to Lord Dalhousie, "under the rule and control of Mr. Ratchford who is everything—keeping a shop for every sort of supply, the whole population is individually indebted to him; selling his goods at enormous profit he makes money of those who pay their accounts, and of those that do not he takes mortgages on their lands, takes that in payment."

The party continued the journey on horseback along the road running from Parrsboro up the valley, where settlement was already well established. They started early in the morning and stopped for breakfast about ten miles inland at an inn run by a magistrate named Jesse Lewis, the son of Caleb Lewis, one of the Loyalist pioneers in the area. Caleb Lewis had come to the province at the end of the war and established himself on a thousand-acre grant of which he had cleared two hundred acres for the farm where his son managed the inn. Here he brought his wife and raised a large family, who remained in the area. By 1818 the population had grown to about three hundred people related to him "by blood or marriage."

We hear nothing of the rest of Lord Dalhousie's journey, except that the road was good and the scenery pleasing and they had covered the thirty-four miles from Parrsboro to Amherst by 2:00 PM. Early maps show that the Main Post Road from Parrsboro ran on the west side of Farrells River to Halfway River. It then turned northeastwards to follow the Maccan River before crossing the bridge over the Nappan River for the final approach to Amherst. Lodging in Amherst was provided by a Mr. White, and the following day, after an expedition to Baie Verte, Lord Dalhousie stayed at Fort Cumberland on the New Brunswick side of the border.

On their return journey, the party took the overland route through Londonderry to Onslow and returned to the capital by way of Truro. The original overland route from Amherst ran through Nappan, River Philip, and Londonderry. It had gradually grown from a mere footpath with successive widening and clearing to the point that by 1809 the mail could be transported by cart rather than by a mail carrier on horseback. When Lord Dalhousie rode over it in the summer of 1818, he pronounced the road from Amherst to River Philip to be "excellent, and the woods magnificent all the way." His party stayed at Phillymore's Inn on the River Philip before embarking on the next stage, across the Westchester Mountain to Sutherland's Inn where they breakfasted (Sutherland's Lake retains the name) and on to Londonderry. The road here was equally pleasing, "green turf with very noble woods all the way along—hilly till within four miles of Flamingo in Londonderry it is flat and dry gravel." Despite the grand scenery, it was not until 1842 that a coach service was possible on this route. Today, a stretch of the older road runs parallel to Highway 104 through the Westchester area.

Cape Breton Roads

European settlement on Cape Breton Island began in the seventeenth century with Nicolas Denys's fishing stations at St. Peters and St. Anns, and continued into the eighteenth century with the establishment of the Fortress of Louisbourg. There were also subsidiary settlements at Port Dauphin (Englishtown) and Boularderie Island, as well as scattered farms elsewhere. This meant that the French presence was divided between the northern and southern areas of the island. Communication between the two was mostly by water, through the Bras d'Or Lake or by sea. The French built only one important strategic road over which supplies that had been carried by ship through the Bras d'Or Lake were

Part of MacKay's A New Map of Nova Scotia *shows Cape Breton's chief roads in the 1820s, including the "Old French Road" to Louisbourg.*

carried to Louisbourg. It ran from East Bay with, presumably, a ferry across the Mira River as no bridge is shown in the 1751 map on which it appears. With the fall of the fortress the road lost its importance. It appears again on William MacKay's *New Map of Nova Scotia* (published in 1829), still in use and identified as "Old French Road." Much of that road has vanished long since, and only a trail still runs for a few miles from the head of East Bay. A section can still be followed east of the Mira, where part of Route 327 passes through the community still known as French Road, and a minor road continues eastwards towards the Louisbourg park site.

When the British took possession of Cape Breton Island, some French settlers remained to be joined by new immigrants, mostly Highland Scots. The centres of population were widely scattered, chiefly around the Strait of Canso and in the Sydney area. Between 1784 and 1820, the island was a separate colony, subject to general neglect by the British, and road building was not a high priority. Coastal communities and those on the Bras d'Or Lake could share communications by water. As a result, when Cape Breton was reintegrated with Nova Scotia in 1820, the roads were still in a very primitive state. Sir James Kempt, lieutenant-governor of Nova Scotia for the period after re-annexation, was keen to bring them up to a reasonable standard, but the task was daunting and money was not always easy to pry out of the legislative assembly for such a remote project.

The postal services on the island were poor, and mail originally travelled from Halifax to Sydney by sea, to be carried southwards by a courier to St. Peters and Arichat. This route from Sydney, dating back to the early days of settlement, was described only as "a trail of sorts." By the mid-1820s, it appeared on a map with the status of an "intended road." Ten years later, it had become the "Post Road from St. Peters and Halifax" and was thus at last recognized as the official link between the provincial capital and the Town of Sydney.

By the late 1820s, a circuitous, mountainous route led from Sydney to St. Anns and Baddeck. Roads also led to the mining areas of Lingan and Cow Bay (Port Morien). There was a roadway forming a loop around the lakeshore along St. Andrews Channel to Grand Narrows and back along East Bay towards Sydney. In the south, a network of roads developed in both directions from the Strait of Canso, eastward to St. Peters and then to Sydney, and westward along the shore to Port Hood and beyond.

The ferries across the Strait of Canso were an integral part of communications between Cape Breton and the rest of Nova Scotia. In 1819 Hugh MacMillan was operating a ferry between Aulds Cove and Plaster Cove (Port Hastings). Ten years later he asked for an increase in the subsidy provided by the assembly in order to finance the larger boats he needed to transport horses and cattle. By 1847 two more ferries were in service. One plied between Mulgrave and Port Hawkesbury and the other, more convenient for travellers to and from Guysborough, went between Steep Creek and McPherson's Ferry. These ferries also required a "scow" or barge onto which cattle, horses, and carriages could be loaded. Today, the Canso Causeway has replaced the ferries that once linked the island with the mainland.

The West Coast of Cape Breton, 1820s

On William MacKay's 1829 map, roads are shown running from Port Hood to Margaree, from Margaree towards Chéticamp, and from Margaree to Baddeck. There were also routes running inland from both Judique (the River Denys Road) and Port Hood, but no road is shown linking the two, nor is one shown southwards from Port Hood to the Strait. In 1818 Lord Dalhousie observed from the sea that "ranging up along Cape Breton shore to Cape Hood, it appears hilly, an unpenetrated hardwood forest." Nevertheless, a road of sorts was in place all along the coast by 1823, when Judge John George Marshall recorded his travels while exercising his judicial duties. The judge's comments are undoubtedly more reliable than the cartographer's record which, despite MacKay's claim to have used "the latest surveys," may have been out of date by the time his map was published.

Judge Marshall recorded some interesting details of these roads. Between Margaree and Chéticamp in 1823, the road near the shore was passable only on foot, with few inhabitants on its six-mile length. The road along the Margaree River he considered "very fair for foot passengers and horses." From Margaree to the head of Broad Cove, the road was "passable on foot and on horseback." Of the twelve miles along the coast to Mabou, he wrote, "may also be passed with a horse, but some of it with a great deal of difficulty especially about the Cape of Mabou where the passage is even somewhat perilous owing to the very steep and broken nature of the ground and the depth of the ravines. The chief mode of passing along this coast is by water." From Mabou to Port Hood, the judge considered the road to be "in a very fair state." From Port Hood through the Judiques, where previously land communication had not existed, the road was described as "in very good order so much so that a carriage might very well be driven over it," suggesting that it was a relatively new development, and the next eighteen miles were "in general very fair." Clearly the situation was improving, and Marshall's report to the lieutenant-governor in 1826 showed that where money had been allocated for repairs, better roads were in place. There was still much to be done, money to be expended, and local interests to be appeased:

> *It appears most probable that the inhabitants [near Chimney Corner] in agreeing upon the present line of road felt desirous of yielding as little as possible of their improved or unimprovable land and therefore in many places have left the road too near to the steep banks of the shore and carried their fences so low down that sufficient room is not left for the making of a good road in addition to which the land for the purpose is not so good as it is higher up and the present line is more circuitous than it need to be.*

From the Strait of Canso to Sydney, ca. 1830

The journey from Halifax to Sydney, undertaken so easily today, was far more daunting in the early years after re-annexation. Upon disembarking from the ferry, the traveller would make his way inland from Ship Harbour, where an old military road that had first been surveyed in 1814 led to the bridge over the Inhabitants River and on to West Bay on the Bras d'Or Lake. It was described by Judge Marshall in 1823 as "very bad, scarcely passable with a horse."

During the 1820s steps were taken to upgrade the existing roads in this area, construct new ones, and replace some of the river ferries with bridges. In 1828 John Munro was sent from Pictou to recommend improvements to the existing road between the Canso Strait and the bridge over the Inhabitants River at what is today Cleveland. Munro was also commissioned to survey the route for a road from the bridge to Grand Digue (now Grandique Ferry). Kempt Road, named after the lieutenant-governor under whose authority this road was built, subsequently became part of Trunk 4. From Grand Digue, a ferry over Lennox Passage took travellers to Isle Madame, while those bound for Sydney continued to St. Peters and from there along the Bras d'Or Lake.

As a result of these improvements, by 1831 the road from the strait to Sydney was considered reasonably passable on horseback. In that year, Munro planned further alterations to the section of road leading inland from Ship Harbour, which, like so many early roads, had originally been laid out in a straight line without regard for the hills. The twelve alterations he designed were completed the following year, and further upgrading of the road between Ship Harbour and Arichat also took place. Within twenty years or so of the re-annexation, the beginnings of a viable road system were taking shape.

Roads around Sydney, 1826-36

In 1826 Judge Marshall had sent a report to Governor Kempt describing the state of the roads around Sydney. His account of the slow progress of work on the road from St. Peters was not encouraging: "The money for this division of the road in the last year was chiefly expended from Curries about six miles onwards which part is levelled and in fair state. The rest appears to be very little improved since the preceding year." The account of the southern section of the road was even gloomier, and at Soldiers Cove "the Mail Courier and other travellers frequently find much difficulty in passing." By 1836 the road was graced with the title of "Post Road from St. Peters and Halifax," offering an overland route from the provincial capital to Sydney.

A network of roads around Sydney linked the mining and fishing areas with its harbour. By 1826 Judge Marshall wrote that the Mira Road was "in tolerably fair shape" but required an alteration to avoid a steep hill; "from Sydney to Lingan, said to be considerably improved by the last sum of money expended on it;" and "from Sydney to Cow Bay, a great part much improved...part still requiring to be fully opened and repaired."

The road from Sydney to Baddeck was particularly difficult because of the mountains. There were also two bodies of water to cross by ferry, St. Andrews Channel and Great Bras d'Or. The judge observed that much repair was required to the road from Little Bras d'Or and across Boularderie Island and "the part between the Great Bras d'Or and St. Annes [is] so very mountainous that it is said to be almost impossible to make it a good road." Early nineteenth-century engineers did not have equipment capable of constructing a road like the present-day one over Kellys Mountain. An alternative route was by way of the "lower road to St. Annes" across Boularderie Island and the ferry to Big Harbour, "this said to be a route very frequently followed by persons travelling between St. Annes and Sydney by reason of the other road...being so very mountainous. It requires to be opened in several places and to be repaired both between St. Annes and Big Harbour, and down the side of the Isle Boularderie to the entrance of the...road across the Island."

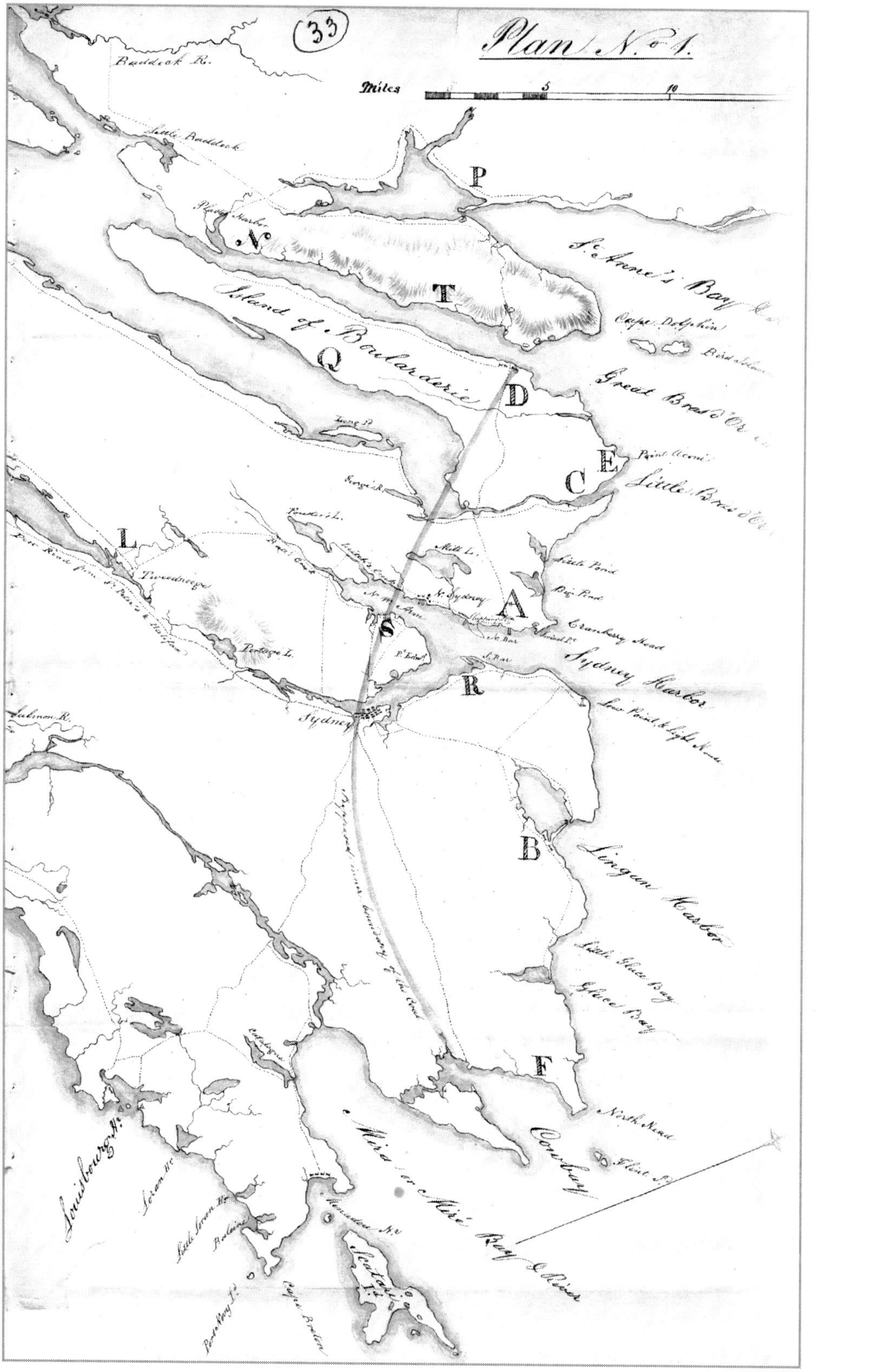
(33)
Plan No. 1.
Miles
5
10
Baddeck R.
Little Baddeck
P
N
T
Q
St. Anne's Bay
Cape Dolphin
Island of Boulardarie
D
Great Bras d'Or
E
Point Aconi
C
Little Bras d'Or
L
Tweedmooge
Mill L.
Little Pond
Big Pond
N. Sydney
A
S
Cranberry Head
Portage L.
Sydney Harbor
R
Sydney
Low Point & Light House
B
Lingan Harbor
Little Glace Bay
Glace Bay
F
North Head
Cowbay
Mira or Miré Bay & River
Louisbourg Hr.
Scatari Id.
Cape Breton

In 1836 both roads leading from Sydney to Baddeck by way of Boularderie Island were still in use, with ferries across St. Andrews Channel and Great Bras d'Or. Today, bridges have replaced the ferries, and Highways 125 and 105 have replaced the old roads across Boularderie Island and over the mountains to St. Annes and Baddeck. The community of Ross Ferry still exists on the shore road along Great Bras d'Or, though a ferry no longer provides a link to the road from Big Harbour to Baddeck.

In Cape Breton today, Trunk 4 still leaves Port Hawkesbury by the old route, crossing the River Inhabitants at Cleveland and turning to take the Kempt Road route to St. Peters. For most of this distance, Highway 104 carries much of the traffic by a route closer to the shore. Trunk 4 continues along the Bras d'Or Lake, following the route of the old Post Road to Sydney. A short section between Portage and Sydney Forks has been straightened, along with other minor alterations. The Trans-Canada Highway (105) on the western side of the lake is now the major route to the Cape Breton industrial area from the mainland, leaving Trunk 4 and the eastern side mainly to the tourists. Here, some of the older roads are still in place, and today's travellers with time to spare can take a more leisurely route through the communities that developed along them.

Northern Nova Scotia and Cape Breton were in many respects the most difficult areas in which to establish a good road system. Both areas were relatively sparsely populated, and each presented major physical obstacles to highway construction and maintenance. Amherst, near the northern boundary of Nova Scotia, was separated from the capital both by distance and by the Cobequid Mountains. Parts of Cape Breton were even more mountainous, while the land route from Halifax to Sydney, at the province's eastern extremity, depended on ferry service across the Strait of Canso. Even after re-annexation to Nova Scotia, the difficulties in travelling the Cape Breton region were compounded by the late start on serious road building and inadequate funding. Nevertheless, communications networks were eventually established in both of these areas that laid the foundation for the travel routes we use today.

Anon. "Plans with Mr. Crawley's Letter...," 1836 (NSARM F/201 1836) This map identifies the Post Road and other roads in the Sydney area.

EPILOGUE

THE EARLY HIGHWAYS OF NOVA SCOTIA are often forgotten today, buried beneath newer roads that follow the same path or lost in the wilderness after replacement by more efficient routes. Some are visible only to the trained eyes of archaeologists, while others survive in different forms: trails, logging roads, or minor dirt roads that are still in local use. These are sometimes identified by signage and local tradition, but often they are less obvious without the help of historical documentation. Their story reminds us of the efforts of early surveying crews and road-builders who defied rocks, swamps, mountains, and dense forests to lay out and construct our roads. They covered seemingly endless distances in all kinds of weather, enduring blackflies, mosquitoes, and other hardships to lay the foundations of the Nova Scotia road system we know today.

Attempts at physically tracing "lost" roads have been made from time to time. As a centennial project in 1967, a group from the University Naval Training Division (UNTD) explored and cut out abandoned sections of the Old Annapolis Road, allowing a group of sea cadets to march over it from end to end that summer. In 1971 members of the History Club at the Sidney Stephen High School in Bedford, with the guidance of Robert Harvey, traced the route of the Old Cobequid Road from Fultz's Corner to Fletcher Bridge. Part of the original road southwest of the Gays River Bridge has recently been traced by archaeologist Sara Beanlands. Probably other groups have worked on similar projects in their own areas.

The discovery of old roads does not have to be a major undertaking. As we drive along Nova Scotia's paved roads, we see clues that alert us to the existence of a former, slightly different route. It can take the form of a highway sign, incorporating names like Post Road, Old Halifax (or the name of any other town or city) Road, Back Road, or the words "diversion" or "loop." Where modern construction has enabled a twisting road to be straightened, older sections often remain to the side. They resemble the oxbows at the side of a river that has cut through meandering banks to find a straighter course. Once people had established homes and businesses on such a road, it was frequently maintained, though probably not much upgraded, even when the main road bypassed it. Or reversing the process, a trunk road may now curve away from the earlier route to serve a community that developed after the original road had been laid out, leaving the straight section as a back road used only as a shortcut. Even while driving a 100-Series Highway today, we can see disused roads that lead nowhere, stopping short at the edge of the highway.

Many of these vestiges are now dirt roads or logging roads, and it is not difficult to imagine them in early days with pedestrians, horsemen, ox- or horse-drawn wagons, buggies, and the occasional more elegant carriage passing through. Before the stagecoaches carried mail to distant parts of the province, the mail carrier on horseback with his saddlebags of packages and letters would be a regular road-user. We can visualize the sleighs that would pass over the roads in winter. We can also imagine their state in wet weather! Some former roads are walking trails today, while others are quite overgrown or have vanished under agricultural, commercial, or residential development. Enough survives to give observant travellers a glimpse into Nova Scotia history dating back to the early nineteenth century.

Acknowledgements

I am indebted to Marie Elwood for launching me on this exploration of the old highways of Nova Scotia and for accompanying me on some early expeditions to trace their routes; to the staff at Nova Scotia Archives and Records Management for their continued patience in locating maps, and particularly to Anjali Vohra for scanning the images used in this book; to Scott Robson at the Nova Scotia Museum, who first identified the Woolford road book, for his help in acquiring images from the museum's collection; to Dianne O'Neill and Shannon Parker at the Art Gallery of Nova Scotia; and to Karen Smith at Dalhousie's Killam Library, Special Collections. Thanks are also due to Sara Beanlands for sharing information about the old Truro Road; to Patrick Murphy at Nimbus for his faith in this project; to Paula Sarson for her sympathetic editing of the text; and to travelling companions Connie Cannon, Nita Graham, and Edith Wolter, who made looking for old roads so much more fun.

Bibliography

Akins, Thomas B. "History of Halifax City." *Collections of the Nova Scotia Historical Society,* vol. VIII, 1892–94. Halifax: Morning Herald, 1895.

Belcher, C. H. *The Farmer's Almanack* (later *Belcher's Farmer's Almanack*) *for the Year of Our Lord ...* Halifax: 1824–1842.

Bird, W. R. "History of the Highways of Nova Scotia," (Unpublished paper, Nova Scotia Historical Society) NSARM, MG 20, vol. 670, no.18.

Black, Mary. "Cape Breton's Early Roads." *The Nova Scotia Historical Quarterly,* vol.5, no 3, (Sept. 1975): pp.277–295.

Calnek, W. A. *History of the County of Annapolis, Nova Scotia: Including Port Royal and Acadia.* Edited and completed by A.W. Savary. Toronto: W. Briggs, 1897.

Campbell, Duncan. *Nova Scotia, in its Historical, Mercantile and Industrial Relations.* Montreal: Lovell, 1873.

"Come to Lunenburg." *Times,* Halifax, July 29, 1845, p.237.

"Conveyance to Windsor" (advertisement). *Acadian Recorder,* Halifax, May 17, 1817.

Cuthbertson, Brian C. *The Loyalist Governor: Biography of Sir John Wentworth.* Halifax: Petheric Press, 1983.

———. *The Old Attorney General: A Biography of Richard John Uniacke 1753-1830.* Halifax: Nimbus Publishing, 1980.

Cuthbertson, Brian C. and Gillis Architects. "Joseph Scott and the Scott Manor House." *Research Papers Prepared for Halifax Regional Municipality,* Halifax, 2002. http://www.halifax.ca/cah/documents/HistoryScottManorHouse.pdf

Dictionary of Canadian Biography Online. http://www.biographi.ca

Dunlap, William. *A History of the American Theatre from its Origins to 1832.* Urbana: University of Illinois Press, 2005.

Evans, R. D. "Stage Coaches in Nova Scotia, 1815 to 1867." Nova Scotia Historical Society, *Collections,* vol. 24, 1938: pp. [107]-134.

"Ezra Witter" (advertisement). *Acadian Recorder,* Halifax, May 17, 1817.

Fingard, Judith, Janet Guildford and David Sutherland. *Halifax: the First 250 Years.* Halifax: Formac, 1999.

Friends of the DesBrisay Museum and Bridgewater Heritage and Historical Society. *One Hundred Years; a Pictorial History of Bridgewater.* Bridgewater: Friends of the DesBrisay Museum and Bridgewater Heritage and Historical Society, 1999.

Graham, Douglas J. "Archibald Family of New Hampshire and Nova Scotia." *A Family Reunion,* 2005. http://home.earthlink.net/~douglasjgraham/Archibal.htm

Halliburton, W. H. O. "Report on the state of the roads in the Middle District of Windsor, 1828." NSARM, RG 1, vol. 235, doc. 3.

Harvey, Robert P. *Historic Sackville.* Halifax: Nimbus Publishing, 2002.

History of Nova Scotia with special attention given to Communications and Transportation. Chapters 6, 7, 8. http://www.alts.net/ns1625/nshist06.html

Howe, Joseph. *Western and Eastern Rambles.* Edited by M. G. Parks. Toronto: University of Toronto Press, 1973.

Jordan, Maureena. *Genealogy of the Jordan/Jordain Family of Guysborough County, Nova Scotia.* http://freepages.genealogy.rootsweb.ancestry.com/~jordain/

Lockwood, Anthony. *A Brief Description of Nova Scotia.* London: Hayden, 1818.

———. Advertisement for the sale of his property, *Acadian Recorder*, August 8, 1818.

Moorsom, William Scarth. *Letters from Nova Scotia Comprising Sketches of a Young Country.* London: Colburn & Bentley, 1830.

———. *Letters from Nova Scotia*, edited by Marjorie Whitelaw. Ottawa: Oberon, 1986. (Includes records by other travellers.)

More, James F. *The History of Queens County, N.S.* Halifax: Nova Scotia Printing Co., 1873.

"New Road to Chester." *Nova Scotian* (Halifax, 1845): p. 330.

Notes on the Oldmixon Family, NSARM, MG 1 vol. 2433, no. 3-4.

Parks, Malcolm G. *The Rising Village: Chapters in the History of Petite Riviere, Nova Scotia.* Lockeport: Community Books, 1998.

Patterson, George. *A History of the County of Pictou, Nova Scotia.* Montreal: Dawson, 1877.

Piers, H. *Evolution of the Halifax Fortress, 1749-1928.* Revised, edited, and completed by G. M Self, Phyllis Blakeley. Halifax: Public Archives of Nova Scotia, 1947.

———. [Notes about inns on the Windsor Road] NSARM MG 1, vol. 1 464, no. 123.

Public Archives of Nova Scotia. *Places and Placenames of Nova Scotia.* Introduction by C. B. Fergusson. Halifax: Public Archives of Nova Scotia, 1967.

Ross, Sally, and Alphonse Deveau. *The Acadians of Nova Scotia Past and Present.* Halifax: Nimbus Publishing, 1992.

Stone, Arthur. *Journey through a Cape Breton County*. Sydney: Cape Breton University Press, 1991.

"The Old Coach Road (1761-1847)." *Firsts of Cumberland County's* [*sic*]. http://www.rootsweb.ancestry.com/~nscumber/files/webpages/CumberlandCountyFirsts.html

The Nova Scotia Atlas, 5th edition. Halifax: Formac, 2001.

"Tracadie." *Antigonish County Placenames*. http://www.parl.ns.ca/placenames/antigonisht.html

University Naval Training Division Centennial Project, May 29–Sept. 17, 1867. NSARM, MG 100, vol. 214, no.9.

Whitelaw, M. ed. *The Dalhousie Journals,* vol.1. Ottawa: Oberon Press, 1978.

Withrow, Alfreda. *St. Margaret's Bay: An Historical Album*. Halifax: Nimbus Publishing/Four East, 1997.

Image Credits

Numbers refer to the page number on which image(s) appear.

Woolford's "Surveys" map courtesy Nova Scotia Archives and Records Management Map Collection (location 15.1).

Other Nova Scotia Archives and Records Management (NSARM) images:

iv-v, 114, 117: NSARM Map Collection (location 3.5.6.1829)

2: NSARM Map Collection, REO, Y7

76, 78, 79: NSARM Map Collection (location 4.1.3.1)

83, 85: NSARM Map Collection, 209/Roads-Hants Co.-Kings Co.-1800 (loca tion 4.1.2.1)

84, 103: Captain W. Moorsom, *Letters from Nova Scotia; comprising sketches of a new county*

88: Nova Scotia Crown Lands, RG 20, Series C, vol. 85, no. 152

96: NSARM Map Collection (location 4.1.1.10)

99, 100: NSARM Map Collection (location 17.2.10)

102: NSARM Map Collection, F/209-Sydney County-undated

106-107: NSARM Map Collection, F/209-Sydney County-Halifax County-wm 1819

108, 110: NSARM Map Collection, 209/Roads-Guysborough County-1840 (location 4.1.2.11)

122: NSARM Map Collection F/201-Cape Breton-1836

History Collection, Nova Scotia Museum: 23, 39, 44, 48, 54, 60, 93, 94.

William Inglis Morse Collection, Special Collections, Dalhousie University Libraries: 74.

Joan Dawson: 9, 28, 29, 31, 75, 97.

Collection of the Art Gallery of Nova Scotia:
30: Captain Michael Seymour, *Fultz's Inn, 12 Miles from Halifax, 1846*
Watercolour and pencil on paper, 18.0 x 25.6 cm

Purchased with funds from the Government of Canada under the terms of the Cultural Property Export and Import Act, Ottawa, and Templeton Management Inc., Toronto, 1998. (1998.306)

81: Mary G. Hall, *View of Digby, - Nova Scotia, 1835*
Lithograph on paper, 25.1 x 34.6 cm

Collection of the Art Gallery of Nova Scotia. Purchased with funds provided by the Government of Canada under the terms of the Cultural Property Export and Import Act, and by John and Norma Oyler, Halifax, 2002. (2002.214.5)

86: William Eagar, *Windsor, Nova Scotia, from Fort Hill, 1840*
Lithograph on chine collé, 16.1 x 25.7 cm

Gift of John and Norma Oyler, Halifax, Nova Scotia, 1995. (1995.85.1)

90: Captain Michael Seymour, *From the Preston Road near Dartmouth, October 15th, 1845*
Watercolour on paper, 17.5 x 25.1 cm

Purchased with funds from the Government of Canada under the terms of the Cultural Property Export and Import Act, Ottawa, and Templeton Management Inc., Toronto, 1998. (1998.304)